Tribology in Vehicles

Tribology in Vehicles

Guest Editors

Shuwen Wang
John Williams
David Charles Barton
Chunxing Gu

MDPI

Basel • Beijing • Wuhan • Barcelona • Belgrade • Novi Sad • Cluj • Manchester

Guest Editors
Shuwen Wang
School of Mechanical Engineering
University of Shanghai for Science and Technology
Shanghai
China

John Williams
Engineering Department
University of Cambridge
Cambridge
UK

David Charles Barton
School of Mechanical Engineering
University of Leeds
Leeds
UK

Chunxing Gu
School of Mechanical Engineering
University of Shanghai for Science and Technology
Shanghai
China

Editorial Office
MDPI AG
Grosspeteranlage 5
4052 Basel, Switzerland

This is a reprint of the Special Issue, published open access by the journal *Lubricants* (ISSN 2075-4442), freely accessible at: https://www.mdpi.com/journal/lubricants/special_issues/1WY7P485V4.

For citation purposes, cite each article independently as indicated on the article page online and as indicated below:

Lastname, A.A.; Lastname, B.B. Article Title. *Journal Name* **Year**, *Volume Number*, Page Range.

ISBN 978-3-7258-6107-1 (Hbk)
ISBN 978-3-7258-6108-8 (PDF)
https://doi.org/10.3390/books978-3-7258-6108-8

Contents

About the Editors

Shuwen Wang

Shuwen Wang is a professor of mechanical engineering at the University of Shanghai for Science and Technology (USST) and a visiting professor at the University of Leeds. He received his PhD from the University of Cambridge, United Kingdom, and his master's degree from the University of Toronto, Canada. His current research interests include tribology, surface science and engineering, structural dynamics, vehicle NVH and tribology, bio-inspired surface design, and laser surface modification. He is a member of American Society of Mechanical Engineers, SAE International, Institute of Physics and Society of Tribologists and Lubrication Engineers.

John Williams

John Williams is Emeritus Professor of engineering tribology at the University of Cambridge where, after graduating in Mechanical Sciences, he completed his PhD in the Department of Physics supervised by Professor David Tabor. He worked for a number of years for Tube Investments on a variety of industrial projects before returning to academia first at the University of Warwick and then in Cambridge. He is the author of the textbook 'Engineering Tribology' and of numerous journal and conference papers. He is a fellow of the Institution of Mechanical Engineers and of the American Society of Mechanical Engineers. He was co-director of the successful Cambridge Tribology Course which ran annually from 1993 to 2019. He is a winner of the Tribology Trust Silver Medal and of James Clayton and Donald Julius Groen Prizes from the Institution of Mechanical Engineers and, in 2005, was elected to the Royal Academy of Engineering.

David Charles Barton

David Barton is currently Professor of Solid Mechanics in the School of Mechanical Engineering at Leeds University, UK. He is a Chartered Engineer, Fellow of the UK Institution of Mechanical Engineers and Fellow of the UK Institute of Physics. Prior to joining Leeds as a lecturer in 1985, Professor Barton had five years experience as a structural analyst in the nuclear power industry mainly concerned with high temperature and aseismic assessment of reactor structures. His research interests at Leeds have focused on the experimental derivation and numerical implementation of complex models of material and structural behaviour, particularly in relation to high strain rate deformation (including crashworthiness of vehicles), tribological interfaces in automotive engineering (brakes, gearbox synchronisers) and biomedical applications (total artificial joints, impact biomechanics).

Chunxing Gu

Chunxing Gu is an Associate Professor and Master's Supervisor in the School of Mechanical Engineering at the University of Shanghai for Science and Technology, where he leads research in tribology and surface engineering. He completed his PhD in Mechanical Engineering at Shanghai Jiao Tong University in 2018 before joining the academic staff at USST. His research encompasses multi-scale lubrication, contact mechanics, wear mechanisms, and surface texture optimization. He has secured several major research grants as principal investigator, including projects from the National Natural Science Foundation of China and the Shanghai Science and Technology Commission, while also contributing to other collaborative research initiatives. His scholarly work has been recognized through awards such as the Shanghai Young Tech Talent Sailing Scholar, and he serves as a reviewer for several leading international tribology journals.

Preface

It is self-evident that all road and rail vehicles, regardless of type of powertrain, rely on components and systems which involve the relative motion of surfaces in close contact with one another. Examples include piston ring contacts within an internal combustion engine, meshing gears, journal bearings and seals, railway wheels on steel rails, and friction brakes. The relative motion in such components gives rise to tribological effects such as friction, wear and noise/vibration which can increase energy consumption and limit the service life of components. Many of these problems can be alleviated by some form of solid or liquid lubrication applied to the contacting surfaces. Hence this Special Issue on the theme of *Tribology in Vehicles* is very relevant to its parent journal *Lubrication*.

This first volume of the Special Issue reflects the range and diversity of tribological issues encountered in modern road and rail vehicles. Thus the volume includes papers on conventional topics such as lubricant developments for internal combustion engines, the contact dynamics in a conical surface wet clutch, and the wear of railway wagon wheels. Also included in the volume are papers specific to bearings and seals in electric vehicles. Finally, problems of friction prediction and squeal vibration in dry friction brakes are addressed in two further papers.

It is hoped that readers will find all these papers of interest and relevant to their own research. We encourage readers to submit original papers to future special issues in this series.

Shuwen Wang, John Williams, David Charles Barton, and Chunxing Gu
Guest Editors

lubricants

MDPI

Article

Braking Friction Coefficient Prediction Using PSO–GRU Algorithm Based on Braking Dynamometer Testing

Shuwen Wang [1,*], Yang Yu [1], Shuangxia Liu [1] and David Barton [2]

1 College of Mechanical Engineering, University of Shanghai for Science and Technology, Shanghai 200093, China
2 School of Mechanical Engineering, University of Leeds, Leeds LS2 9JT, UK; d.c.barton@leeds.ac.uk
* Correspondence: shuwenwang66@163.com

Abstract: The coefficients of friction (COFs) is one of the most important parameters used to evaluate the braking performance of a friction brake. Many indicators that affect the safety and comfort of automobiles are associated with brake COFs. The manufacturers of friction brakes and their components are required to spend huge amounts of time and money to carry out experimental tests to ensure the COFs of a newly developed braking system meet the required standards. In order to save time and costs for the development of new friction brake applications, the GRU (Gate Recurrent Unit) algorithm optimized by the improved PSO (particle swarm optimization) global optimization method is employed in this work to predict brake COFs based on existing experimental data obtained from friction braking dynamometer tests. Compared with the LSTM (Long Short-Term Memory) method, the GRU algorithm optimized by PSO avoids the accuracy reduction problem caused by gradient descent in the training process and hence reduces the prediction error and computational cost. The combined PSO–GRU algorithm increases the coefficient of determination (R^2) of the prediction by 4.7%, reduces the MAE (mean absolute error) by 14.3%, and increases the prediction speed by 40.1% compared with the standalone GRU method. The prediction method based on machine learning proposed in this study can not only be applied to the prediction of automobile braking COFs but also for other frictional system problems, such as the prediction of braking noise and the friction of various bearing transmission components.

Keywords: friction; machine learning; RNN; optimization; PSO; disk brake

1. Introduction

Friction braking systems are critical to the safety, handling stability, comfort, and performance of all automobiles, including electric vehicles, which, despite the use of regenerative braking, are still required to carry friction brakes. The coefficients of friction (COFs) at the brake sliding interface determine the quality of the braking performance of a vehicle and play an important role in the generation of braking noise and vibrations. Since the requirements of braking COFs imposed by an automobile's original equipment manufacturers (OEMs) are very strict, braking system manufacturers need to spend a lot of human resources, testing, and financial resources to ensure that the braking COFs meet the requirements of the OEMs. If a vehicle is running at a high speed and the braking COFs are too small, the braking forces could be insufficient to decelerate the vehicle safely, which may cause serious traffic accidents. On the other hand, if the braking COFs are too large, the road wheels of the vehicle could become locked during braking operation, causing the vehicle to flick its tail, slip sideways, or even roll over. High COFs are also associated with increased noise and vibrations from a friction brake, which can lead to high warranty costs for the vehicle manufacturer. Therefore, if the braking COFs can be quickly and accurately predicted and the braking noise can be assessed according to the predicted COFs, the time spent on the development of new braking systems can be shortened, which can significantly improve the economic efficiency and profitability of brake manufacturers.

Although the generation mechanism and control techniques of frictional braking noise and vibrations have been studied for nearly one hundred years, the accurate prediction and control of friction-induced vibrations and noise have always been a challenge in the design of braking systems [1,2] due to the lack of the knowledge of real-time dynamic friction between the brake disk and pads during braking. Therefore, the prediction of friction in braking has attracted many researchers since the 1930s. Khairnar et al. [3] computed the COFs for symmetric and asymmetric drum brake shoes. The extracted COFs were used in an antilock braking system (ABS) algorithm to calculate the brake torque. Riva et al. [4] used a finite element analysis (FEA) approach combined with a coefficient of friction (COF) p-v map to compute the global COFs of a disk brake system. The local COFs were determined from a p-v map for each local sliding velocity and the contact pressure was determined by the FEA. Based on the local COFs, the braking force of the entire brake system and the global COFs can be evaluated. Meng et al. [5] reviewed a large number of peer-reviewed papers and concluded that COFs were critical to many areas, such as lubrication, wear, surface engineering, etc. In addition, COFs play a very important role in the theories or mechanisms of the generation of frictional vibrations and noise, e.g., in the stick–slip theory [6] and self-lock–slip theory [7]. A large number of studies have proved that the COFs have a significant effect on the generation of frictional vibrations and noise [8–10]. Jarvis et al. [9] believed that the COFs can be used as a good indicator of the tendency of braking friction materials to produce noise. Nishiwaki et al. [10] developed a theoretical model to study drum and disk brake noise and concluded that the braking noise was caused by the dynamic instability of the braking system and the transient change of COFs.

In recent years, due to the rapid development of machine learning (ML) theory and artificial intelligence (AI) technology, more and more ML algorithms have been applied in regression and classification prediction applications. Zhang et al. [11] employed the LSTM (Long Short-term Memory) algorithm to predict operation conditions of industrial IoT equipment and Jiang et al. [12] applied it to predict the health evolution trends of an aero-engine. Zhang et al. [13] combined CNN (Convolutional Neural Networks) and SVM (Support Vector Machine) for fault diagnosis of braking friction of mechanical equipment. Zhang et al. [14] used the CNN algorithm for feature extraction and combined it with the GRU algorithm to predict the uneven wear state of the friction block. Yang et al. [15] applied deep RNN (recurrent neural network) to the dynamic state estimation of the advanced brake system of electric vehicles. Sabanvoic et al. [16] used the lightweight squeeze net deep neural network model for the identification and classification of road-surface types so as to identify road pavement surfaces with different COFs. Stender et al. [17] used Convolutional Neural Networks (CNN) based on machine learning to detect vibrations combined with the RNN algorithm to predict braking noise. Wang et al. [18] used the LSTM algorithm and the optimized XGBoost algorithm to predict the braking COFs and braking noise, and the prediction results were in good agreement with the experimental results. Alexsendric and Barton [19] used artificial neural networks (ANN) to predict the COFs of a disk brake system for different operating conditions taking account of the composition of the friction material. The Bayesian regulation learning algorithm was found to give the best fit to the experimental data. Alexsendric et al. [20] also used ANN techniques coupled with the Bayesian regulation learning algorithm to investigate brake fade and recovery following high-temperature brake operation as a function of the composition and manufacturing process of the brake pads.

The main research goal of this present work is to use the GRU algorithm combined with particle swarm optimization (PSO) to enable the rapid and more accurate prediction of brake COFs so as to realize the fast design of brakes, shorten the experimental period, reduce the design cost, effectively control the quality, and improve the efficiency of braking system development. The overall concept is that if an experimental data set is available over a limited range of braking conditions, the model can be trained to use this data to allow for the prediction of COFs for different conditions, albeit these conditions must be the same rotor/pad combination. Thus, if a manufacturer wants to modify the test conditions for a new application, there is no need to repeat the experimental test program, which is expensive and time-consuming, because reliable predictions of COFs can be obtained from the model.

The overall methodology proposed can be divided into the following steps: first, collecting experimental data via the Link3900 brake dynamometer testing; second, data feature engineering that includes dealing with the missing values and outliers of the experimental data, investigating the correlation between one feature and another and between features and targets, dividing all data sets into training sets and test sets; and finally, using the optimized GRU algorithm to predict the braking COFs. Figure 1 illustrates the research procedure adopted for this study, which also reflects the structure of the paper.

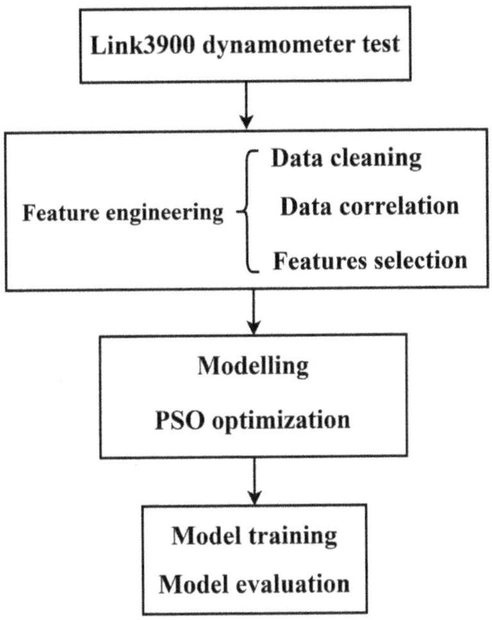

Figure 1. Research procedure and the structure of the paper.

2. Braking Dynamometer Testing and Typical Results

The Link3900 brake performance test bench produced by LINK company (Shanghai, China) was employed for the experimental tests. The test bench is a standard testing machine for testing the performance of various vehicle brakes, as shown in Figure 2. Tests were carried out according to the SAE-J2521 standard [21], which defines a set of experimental procedures recognized by the industry to simulate the braking process of a vehicle. The SAE-J2521 standard braking test procedure consists mainly of the following basic brake conditions: Snub Brake, Brake, Deceleration Brake, Cold Brake, and Fade Brake, which includes a total of 2321 braking stops in 31 different test modules with various braking conditions.

In this present work, nearly 1000 separate items of COF data were obtained from braking tests of one particular friction pair consisting of a grey cast iron brake disk with laser-machined 96 M-shaped grooves on each of its frictional surfaces and standard NAO (non-asbestos organic) friction pads. The COFs obtained under various braking conditions were used for the training and validation of the prediction of braking COFs. More detailed information about the brake materials and the setup of the braking dynamometer testing can be found in Ref. [18]. Table 1 demonstrates typical experimental data obtained from just 11 out of the approximately 1000 braking dynamometer tests. In these 11 tests, the main parameter varied was the average deceleration (Avg Decel) of the simulated braking event, which was from the shown initial speed to a brake release speed of 30 kph.

Figure 2. LINK3900 brake performance test rig (partial).

Table 1. Screenshot of typical experimental data obtained from braking tests.

Stop	Brake Speed (kph)	Release Speed (kph)	Stop Time (s)	Avg Decel (g)	Avg Torq (Nm)	Max Torq (Nm)	Avg Press (bar)	Max Press (bar)	Avg μ Level	Initial Temp Rotor C	Final Temp Rotor C	Peak Level (dBA)	Frequency of Peak (Hz)	Above Threshold 70 dBA
1	79.7	30.0	8.377	0.18	565	617	15.0	15.8	0.44	100	158	73.5	2800	YES
2	80.0	30.0	4.68	0.35	1087	1145	29.9	30.6	0.42	100	163	69.9	5900	
3	80.1	30.0	8.32	0.18	569	620	15.0	15.7	0.44	100	158	59.2	5950	
4	79.8	30.0	7.01	0.22	674	730	18.0	18.6	0.44	100	157	69.0	5975	
5	80.0	30.0	5.99	0.26	819	877	21.9	22.5	0.43	100	158	70.9	2800	YES
6	79.8	30.0	3.83	0.44	1357	1404	37.8	38.6	0.42	100	164	71.5	2800	YES
7	79.7	30.0	8.35	0.18	563	615	15.0	15.9	0.44	100	156	69.1	6200	
8	79.7	30.0	5.25	0.31	948	1010	25.9	26.5	0.43	100	160	69.9	5975	
9	80.1	30.0	7.06	0.22	670	726	18.0	18.5	0.43	100	157	67.6	6325	
10	80.0	30.0	4.19	0.40	1232	1288	33.9	34.6	0.42	100	160	71.4	2925	YES
11	80.0	30.0	8.36	0.18	564	621	15.0	15.8	0.44	100	155	68.6	6350	

3. Feature Engineering

3.1. Data Cleaning

To enable the accurate prediction of braking COFs, two data cleaning methods are used in this study, namely Min–Max scaling [22] and Z-Score standardization [23], as the input and output ranges of the nonlinear activation functions used in the GRU neural network model need to be defined when predicting the experimental data. The two kinds of data cleaning processing are described below.

3.1.1. Min–Max Scaling

Min–Max scaling, also known as deviation normalization, is a linear transformation of the original data. The original value x is normalized by Min–Max to give a standardized value in the interval [0,1], and its calculation formula is shown in Equation (1):

$$X\prime = \frac{x - x_{\min}}{x_{\max} - x_{\min}} \tag{1}$$

where x_{max} is the maximum value of all the data samples, x_{min} is the minimum value of all samples, and X' is the standardized value.

3.1.2. Z-Score Normalization

The Z-Score normalization aims to reduce the amount of calculation and improve the efficiency of the model. Z-Score normalization, also known as standard score, is the difference between a data sample and the average value of all data μ divided by the standard deviation σ of the data set. In most cases, by centralizing and standardizing the data, we will obtain the data in the form of the standard normal distribution (where the mean value is 0 and the standard deviation is 1). The calculation formula is shown in Equation (2):

$$x\prime = \frac{x - \mu}{\sigma} \tag{2}$$

where μ refers to the mean of all samples, σ refers to the standard deviation of all samples, and x' is the normalized result.

3.1.3. Outlier and Missing Value Treatment

In many cases, the original data obtained have outliers and/or missing values; if the abnormal and missing data are not processed, the machine learning algorithm will not be able to work. In this study, the Fillan method programmed in the Python (3.9.16) language is used to automatically detect and fill in the missing values in the data set.

Figure 3 presents the box plot of the experimental data, where the input features of BSpe, RSpe, ADec, ATor, MTor, APre, MPre, ACof, ITem, and FTem represent the initial braking speed, release speed, average deceleration, average torque, max torque, average pressure, max pressure, average COF, initial braking temperature, and final braking temperature, respectively. The main function of the box plots is to check whether there are outliers in the data. If the data appear outside the box, it means that the data are outliers; otherwise, they are normal data.

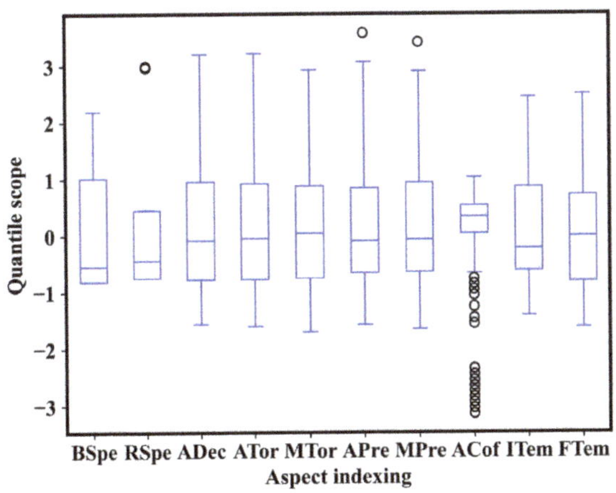

Figure 3. BP box plot.

It can be seen from Figure 3 that there are several abnormal data (circles out of the boxes in Figure 3) in the chosen features of release speed, average pressure, and max pressure, which were eliminated from the data sets because they are clearly outliers. No abnormal data were found for other features. The averaged COF data or circles out of the box were due to the various braking conditions, e.g., the various initial braking temperatures used in the dynamometer testing, and all these ACof data were used in the modeling training.

3.2. *Investigating the Relationships between Features*

In statistics, there are many ways to measure the degree of correlation between two variables. In this study, the correlation between the features chosen to represent the bench experimental data is evaluated by two methods, namely Pearson product moment correlation coefficient (PPMCC) method [24] and the Maximal Information Coefficient (MIC) method [25].

3.2.1. Pearson Correlation Coefficient

The Pearson product moment correlation coefficient (PPMCC) is widely used in the field of data science. PPMCC is defined as the quotient of covariance and standard deviation between two data sets, which can be calculated using Equation (3):

$$\mathrm{r} = \frac{\sum_{i=1}^{n}(x_i - x)(y_i - y)}{\sqrt{\sum_{i=1}^{n}(x_i - \overline{x})^2}\sqrt{\sum_{i=1}^{n}(y_i - \overline{y})^2}} \tag{3}$$

where $\overline{x}, \overline{y}$ is the average value of x and y for n experimental data, respectively. The PPMCC can range from -1 to +1, with values closer to +1 indicating a stronger positive linear correlation. Conversely, the closer the value is to -1, the stronger the negative linear correlation.

The PPMCC values between the ten input features and the output feature of ACof were calculated in turn, and the correlation values are shown in Figure 4. For example, the value of -0.48 in this figure indicates that there is a negative correlation between the initial temperature and the average COF, that is, the higher the initial temperature, the lower the average COF. This is consistent with the known inverse relationship between COF and rotor temperature.

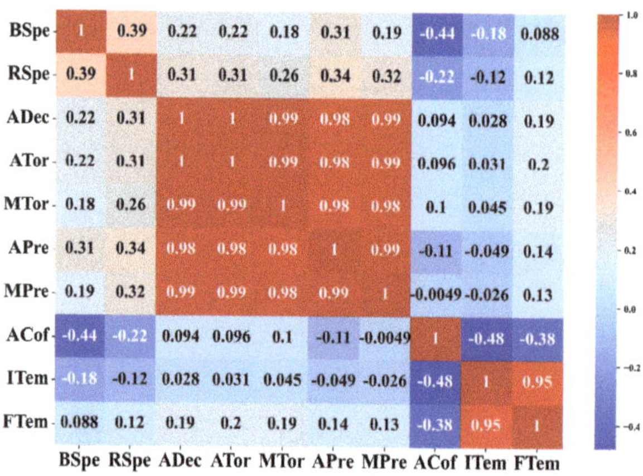

Figure 4. PPMCC matrix heatmap.

3.2.2. Maximal Information Coefficient

The Maximal Information Coefficient (MIC) is used to measure the linear and nonlinear relationships between variables, as well as the non-functional dependence for effective measurement. Let the data set D be a finite ordered data set, which is divided into a grid G, and let $D|_G$ denote the probability distribution of the data set D on the grid G, then the MIC between the variables X and Y may be calculated by Equation (4):

$$MIC(D) = \max_{xy \le B(n)} \frac{I^*(D, x, y)}{\log_2 \min\{x, y\}} \tag{4}$$

where $I^*(D, x, y) = \max I(D|_G)$ is the MIC under different grid division; i.e., for variables on the set D, the X-axis is divided into X grids, and the Y-axis is divided into Y grids of all possible grid G; $I(D|_G)$ is the mutual information given the probability distribution $D|_G$; B is a monotone increasing function and satisfies $B(n) \leq O(n^{1-\varepsilon})$ and $0 < \varepsilon < 1$. In this work, $B(n) = n^{0.6}$ is chosen according to reference.

The MIC can range from 0 to 1, with larger values indicating stronger correlations. Conversely, the smaller the absolute value, the weaker the correlation. The MIC values between the ten input features and the output feature of ACof were calculated in turn, and the correlation values are shown in Figure 5. It can be seen from Figure 5 that the maximum MIC value is 0.52 between features ITem and ACof, indicating that the initial temperature of the disk is the most important factor affecting the braking COFs.

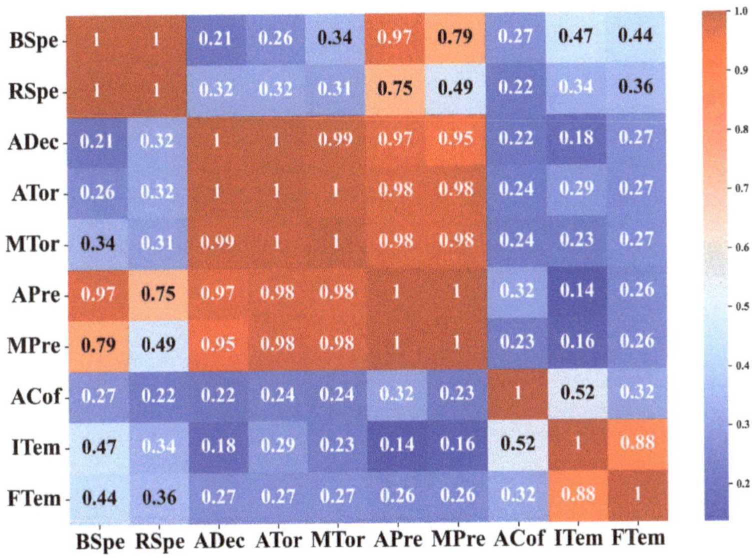

Figure 5. MIC matrix heatmap.

3.3. Selection of Features

Experimental data feature engineering is used to mine the bench experimental data set and investigate internal relationships between the data parameters. Here, it is used to more deeply understand the features affecting braking COFs and remove unnecessary features in order to provide concise and effective model characteristic parameters for the development of a braking COF prediction model. Based on the above PPMCC and MIC analysis results, as shown in Figures 4 and 5, four input features, namely the initial braking speed (BSpe), the brake release speed (RSpe), the average hydraulic pressure (APre), and the initial rotor temperature prior to braking (ITem) were selected to be the most important in order to develop the model for predicting the output feature of the average measured COF (ACof).

4. Prediction Algorithms

The RNN (recurrent neural network) is a kind of recurrent neural network that takes sequence data as the input, re-curses all nodes (re-circulating unit) in the evolution direction of the sequence, and connects them in chains. In deep learning, the RNN is particularly suitable for processing and predicting a class of neural network models related to serial data. Figure 6 demonstrates the structure of a typical RNN cell:

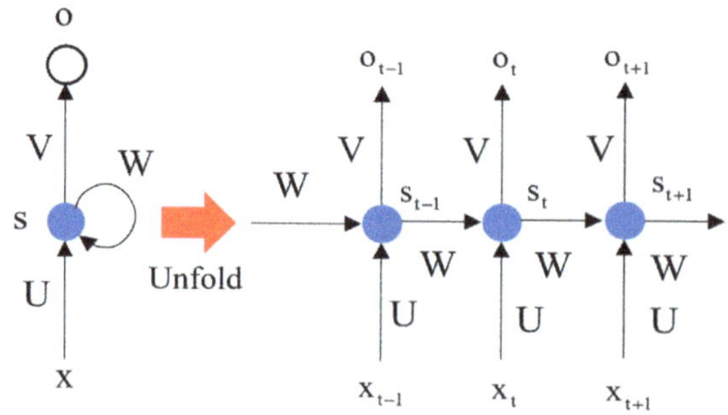

Figure 6. RNN cell expansion structure diagram.

The LSTM (Long Short-Term Memory) and GRU (Gate Recurrent Unit) are two good variants of RNN, which can selectively add or reduce information and can effectively alleviate the problem of RNN gradient disappearance or gradient explosion [26]. In RNN, the gradient is calculated by a multiplication of time steps, which means that the gradient is obtained by the continuous multiplication of multiple identical or similar matrices, such as the cyclic weight matrix. If the values of these matrices are less than 1, the multiplication causes the gradient to decrease exponentially, causing the gradient to disappear. The gradient explosion is similar except, if the values of the matrices are greater than 1, the gradient increases exponentially to very large values.

4.1. LSTM Method

The LSTM model adds input gates, output gates, and forget gates to neurons in each layer corresponding to time points to realize the selective memory of neurons. Its structural principle is shown in Figure 7 (the red X represents multiplication and the blue + represents addition, indicating that the two functions are multiplied and added, respectively), which consists of forget gates, input gates, and output gates. The functions of these three gates are as follows:

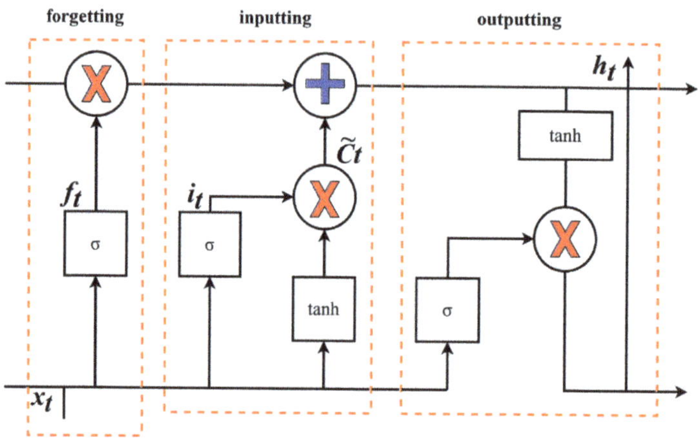

Figure 7. Gated structure in LSTM.

Forget gates are used to decide what information and structure to discard from the "cell". This layer reads the current input x_t and the pre-neuron information h_{t-1}, and it is up to f_t to decide the discarded information.

Input gates are used to determine the new information stored in the cell state to determine the value i_t to be updated; the *tanh* layer is used to create a new vector of candidate values $\tilde{C}_t$ to add to the state.

Output gates are used to determine the value of the next hidden state, which contains the information of the previous input.

4.2. GRU Method

GRU is a new generation of recurrent neural networks. Similar to LSTM, GRU removes cell states and uses hidden states for information transfer. It contains only two gates: the reset gate and the update gate, which are explained as follows:

(1) The reset gate determines how much previous information is forgotten and how new input information is combined with the previous memory, and it uses the current input information to make the hidden state forget any information that is found to be irrelevant to the prediction in the future. It also allows for the construction of more interdependent features. Essentially, the reset gate determines how much of the past data should be forgotten.

(2) The update gate acts similarly to the forget gate and input gate in LSTM. It decides what information to forget and what new information needs to be added. Controlling how much information from previous hidden states gets passed to the current hidden state is very similar to the memory cells in LSTM networks. It helps RNNs to remember long-term information and decide whether to copy all information from the past to reduce the risk of vanishing gradients.

GRU is a very popular network because its structure is simpler and its characteristics are better than the LSTM. Therefore, this study utilizes the GRU algorithm to predict braking COFs.

Gated Computation for GRU

The structure of a GRU neural network is shown in Figure 8, where the red X represents multiplication and the blue + represents addition, indicating that the two functions are multiplied and added, respectively. Compared with the traditional recurrent neural networks, the advantage of the GRU neural network is that it only contains two gates, namely the reset gate and update gate, and the detailed operation of both of which are described below.

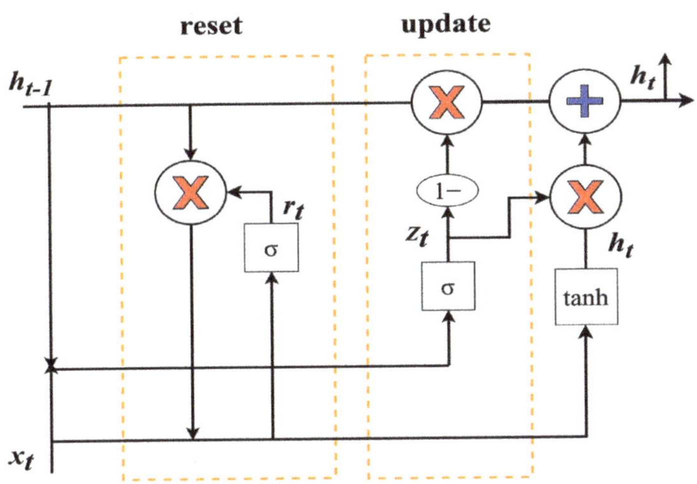

Figure 8. Gated structure in GRU.

(1) Update gate.

At time step t, we first need to compute the update gate Z_t using the following formula:

$$Z_t = \sigma(\omega_Z \cdot [h_t, x_t]) \tag{5}$$

where x_t is the input vector at the t-th time step, that is, the t-th component of the input sequence x, which undergoes a linear transformation when multiplied by the weight matrix ω_Z. The h_t vector holds the information from the previous time step t, which also undergoes the linear transformation. The update gate adds these two pieces of information and feeds them into the Sigmoid activation function, thus compressing the activation result to be between 0 and 1.

(2) Reset gate.

Essentially, the reset gate determines how much of the past information needs to be forgotten, which can be calculated using the following expression:

$$r_t = \sigma(\omega_Z \cdot [h_{t-1}, x_t]) \tag{6}$$

This expression is the same as the expression for the update gate, but the parameters of the linear transformation and its use are different. The h_{t-1} holds the information from the previous time step $t - 1$, which also undergoes the linear transformation.

(3) Current memory content.

Now let us discuss in detail how exactly these gates affect the final output. In the use of reset gates, the new memory content will use the reset gate to store relevant information from the past to calculate the new data:

$$h_t = \tanh(\omega \cdot [r_t \times h_{t-1}, x_t]) \tag{7}$$

The input x_t and the previous time information are first subjected to a linear transformation by the right multiplication matrix w. Then, the Hadamard product of the reset gate r_t with h_{t-1} is computed. Because the reset gate computed earlier is a vector of 0 to 1 values, it will measure the magnitude of the gate opening. For example, if the gating value for an element is 0, it means that the information for that element has been completely forgotten. The Hadamard product will determine the previous information to retain and forget. The results of these two calculations are added together and then put into the hyperbolic tangent activation function.

(4) The final memory of the current time step.

In the final step, the network needs to compute the h_t vector, retain the information from the current cell and pass it on to the next cell. In this process, it needs to use the update gate, which determines what information needs to be collected in the current memory content h_t and the previous time step h_{t-1}. This process can be expressed as:

$$h_t = (1 - z_t) \times h_{t-1} + z_t \times \widetilde{h_t} \tag{8}$$

where z_t is the activation result of the update gate, which also controls the inflow of information in the form of gate control. The Hadamard product of z_t and h_{t-1} represents the information retained in the final memory at the previous time, which, together with the information retained by the current memory in the final memory, equals the output of the final gated loop unit.

4.3. Model Evaluation Index

4.3.1. Correlation Index R^2

R^2 is the determining coefficient, also known as the correlation index, which represents the degree of fit of the regression equation to the data set as a means of measuring the reliability of the prediction of the regression equation. The closer the value is to 1, the closer the actual prediction point is to the goal line, the better the degree of fit of the model to the data, and therefore, the more accurate the prediction of the model. Equation (9) can be used to calculate the R^2 value ranges from 0 to 1:

$$R^2 = 1 - \frac{RSS}{TSS} = 1 - \frac{\sum_{i=1}^{n} (y_i - \hat{y}_i)^2}{\sum_{i=1}^{n} (y_i - \overline{y_i})^2} \quad (9)$$

where RSS is the sum of residual squares, namely:

$$RSS = \sum_{i=1}^{n} (y_i - \hat{y}_i)^2 \quad (10)$$

And TSS is the sum of the squares of the total deviation, i.e.:

$$TSS = \sum_{i=1}^{n} (y_i - \overline{y_i})^2 \quad (11)$$

4.3.2. MAE Index

The mean absolute error (MAE), also known as mean absolute deviation, measures the absolute difference between the real value and the predicted value. The smaller it is, the more accurate the prediction is. MAE is the deviation between all the observed values and the arithmetic mean value, which is obtained by averaging the absolute values. The calculation formula is as follows:

$$MAE = \frac{1}{n} \sum_{i=1}^{n} |h(x_i) - y_i| \quad (12)$$

where $h(x_i)$ is the predicted value and y_i is the measured value.

The MAE avoids the positive and negative errors canceling each other, so it can more accurately reflect the magnitude of the prediction error than the correlation index.

5. Prediction Results

5.1. Model Comparison

The train_test_split function of the sklearn routine within the Python library is used to split the whole dataset into a training set, a validation set, and a test set, accounting for 60%, 20%, and 20% of the available dynamometer data, respectively. The training set is used for model training, the validation set is used for preliminary evaluation of whether the model training results meet the requirements, and the test set is used for model prediction. The final evaluation index of the model was evaluated by comparing the final prediction results with the values of the test set. Then, the pytorch, numpy, and other routines within the Python library are used to define the LSTM and GRU algorithm models, respectively. The model is trained by inputting data representing the independent input features and target output features from the training data set. Predictions are then made using the validation data set composed of a new set of test variables in order to evaluate the model performance and adjust the hyperparameters. The test set data is retained as unique data for the final evaluation of the optimized model performance.

Figure 9 represents the MAE and the R^2 evaluations for the GRU and LSTM predictions for the validation data set, respectively. It is observed from Figure 9 that the MAE of GRU is smaller, while the R^2 value of GRU is larger. The closer the MAE is to 0 and the closer the R^2 value is to 1, the more accurate the prediction. Therefore, the GRU model was selected in preference to the LSTM to further develop the prediction model for the braking COF.

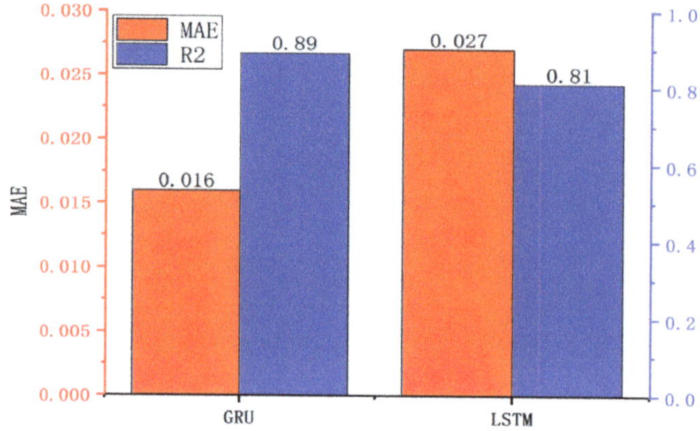

Figure 9. Evaluation metrics.

5.2. Predictions of the GRU Model

Table 2 shows the names and default values of the parameters required for the GRU model. The number of input features has already been set to 4 (BSpe, RSpe, APre, ITem), and the number of output features is 1 (ACof). Based on these parameters, the COF predicted by the GRU algorithm are compared with the validation set of measured data from the bench tests, as shown in Figure 10. It can be seen that, although the general trend of the measured data is well predicted, there are some large discrepancies approaching a maximum of 10% between the predicted and measured COF values. Hence, it was decided to implement the PSO method in order to further optimize the GRU algorithm.

Table 2. Description of the parameters required by the GRU.

Parameter	Parameter Interpretation	Default
input_size	Number of input features	4
output_size	Number of output features	1
rnn_unit	Hidden layers	64
lr	Learning rate	0.001
epoch	Iterations	100

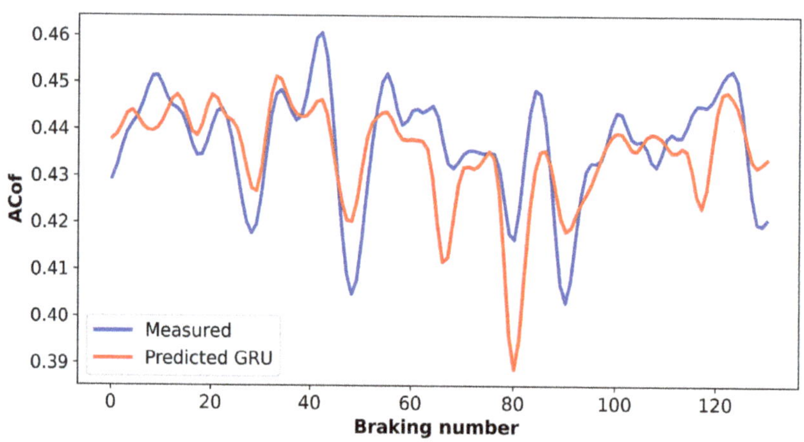

Figure 10. Comparison of predicted and measured COF for test set.

5.3. Predictions of the GRU Model Optimized with the PSO Algorithm

The particle swarm optimization (PSO) algorithm is a global optimization algorithm based on the bionic study of bird foraging behavior in nature. It considers each possible solution in the global variable as a particle that has its own direction and speed so that all particles move towards the local optimal position. By constantly updating the local optimal position *xbest* and the global optimal position *Pbest* of the particle, an optimal solution of the objective function can be obtained.

The hyperparameters of the GRU neural network are classified as the attributes of the particle, and the fitness of the particle is evaluated. Through continuous iteration, the optimal particle is updated. Finally, the optimal particle with the best fitness is obtained to give the required optimal hyperparameters. Figure 11 illustrates the optimization process for the braking COF prediction model using the combined PSO–GRU algorithm.

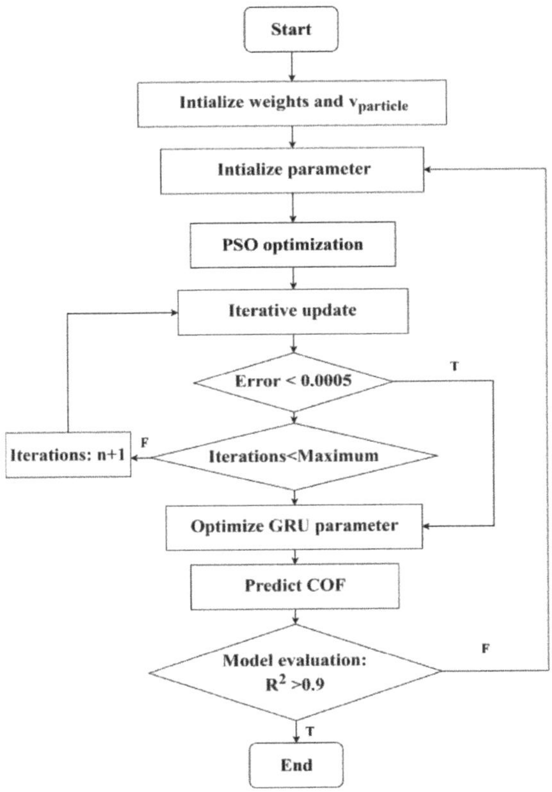

Figure 11. Flow chart of PSO–GRU prediction model for brake COF.

In summary, the PSO is used to optimize the hyperparameters of the GRU algorithm to find the best hyperparameters, and the best hyperparameters are then employed in the GRU algorithm to predict the braking COFs. If the acceptance criterion of $R^2 > 0.9$ is not achieved in the predicted values, the PSO optimization is repeated and the GRU algorithm rerun until the required level of accuracy has been achieved.

The value of the loss function (loss) is a measure of the discrepancy between the predicted value and the measured value. Figure 12 shows that the loss curve of the PSO–GRU model converges faster to a zero value than that of the GRU model used in isolation.

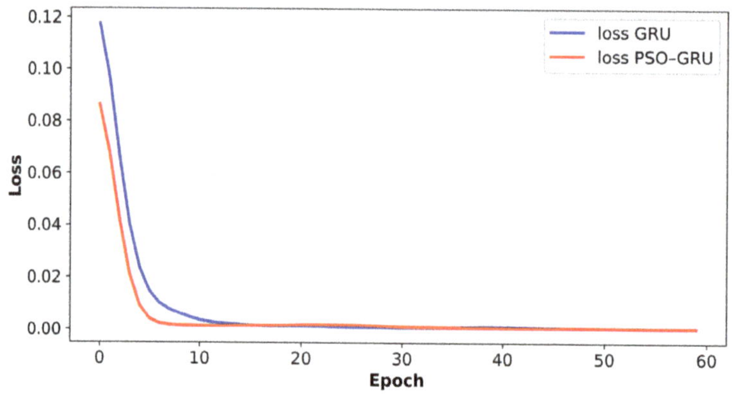

Figure 12. Loss curves of GRU model and PSO–GRU model for test set.

Figure 13 presents the measured COFs and the predicted results from the combined PSO–GRU model for the validation data set. It is obvious from the comparison of Figures 10 and 13 that the PSO–GRU algorithm exhibits superior prediction performance compared to the GRU-only algorithm.

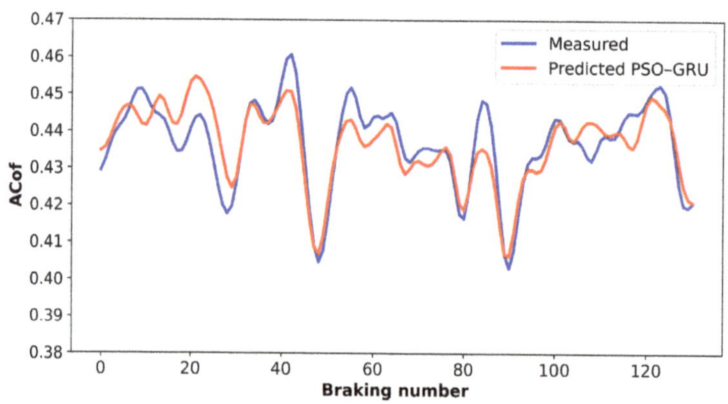

Figure 13. Comparison of the predicted and measured COF for test set after model optimization.

5.4. Model Comparison

The same experimental data were input into the GRU algorithm and PSO–GRU algorithm, and the model evaluation indicators (R^2, MAE, training time) obtained from the GRU algorithm for the validation data set are compared with those from the combined PSO–GRU algorithm in Table 3. Compared with the GRU algorithm before optimization, the R^2 of the combined PSO–GRU algorithm is increased by 4.7%, the MAE is reduced by 14.3%, and the prediction speed is increased by 40.1%.

Table 3. Performance comparison of GRU and LSTM algorithms.

Model	R^2	MAE	Training Time (ms)
GRU	0.893	0.016	220
PSO–GRU	0.935	0.014	157

6. Conclusions

The rapid and effective prediction of COFs is of great significance for the study of braking performance and frictional noise, and also to further understand the factors controlling friction. One of the biggest challenges to this goal is the nonlinearity of the COF, which is affected by a large number of factors.

In this study, a GRU neural network combined with an improved PSO algorithm has been successfully used to solve the complex nonlinear problem of predicting braking COFs from a limited set of experimental data. The PSO algorithm has been used to improve the convergence rate and prediction accuracy of the original GRU neural network. The PSO–GRU algorithm increased the R^2 of the prediction by 4.7%, reduced the MAE by 14.3%, and increased the prediction speed by 40.1%. It is clear from the results that, after applying the PSO parameter optimization algorithm, the GRU prediction model has a shorter training time and better prediction accuracy for an unseen test data set than the standalone GRU model. Therefore, the combined PSO–GRU algorithm is a better choice for the accurate prediction of COFs from a limited set of experimental data.

The prediction of braking COFs with the PSO–GRU algorithm has significance not only for the fast development and evaluation of an automotive braking system, but also provides meaningful reference for solving complicated tribological problems in other applications.

Author Contributions: Conceptualization, S.W.; Methodology, S.W.; Validation, S.W. and Y.Y.; Formal analysis, Y.Y. and S.W.; Investigation, Y.Y. and S.L.; Data curation, S.W. and Y.Y.; Writing—original draft preparation, Y.Y.; Writing—review and editing, S.W. and D.B.; Supervision, S.W. All authors have read and agreed to the published version of the manuscript.

Funding: This research was financially supported by the Science and Technology Committee of Shanghai Municipality (Grant No. 18060502400) and Natural Science Foundation of Shanghai (Grant No. 21ZR1445000).

Data Availability Statement: The data presented in this study are available upon request from the corresponding author.

Conflicts of Interest: The authors declare no conflicts of interest.

Nomenclature

Symbol	Definition	Unit
ACof	average COF	/
ADec	average deceleration	g
APre	average pressure	bar
ATor	average torque	Nm
B	monotone increasing function	/
BSpe	initial braking speed	kph
COF	coefficient of friction	/
COFs	coefficients of friction	/
FTem	final braking temperature	°C
GRU	gated recurrent unit	/
h_t	hidden state at time step t	/
$h(x_i)$	predicted value	/
$I(D\|_G)$	probability distribution	/
$I^*(D, x, y)$	probability distribution under different grid division	
ITem	initial braking temperature	°C
LSTM	long short-term memory	/
MAE	mean absolute error	/
MIC	maximal information coefficient	/
MIC(D)	value of MIC	/
MPre	max pressure	bar
MTor	max torque	Nm
PPMCC	Pearson product moment correlation coefficient	/

Symbol	Definition	Unit
PSO	particle swarm optimization	/
r	value of PPMCC	/
r_t	activation vector of reset gate	/
R^2	evaluation of the model	/
RSpe	release speed	kph
RSS	the sum of residual squares	/
TSS	the sum of the squares of the total deviation	/
vparticle	particle velocity	/
x_{max}	maximum value of all samples	/
x_{min}	minimum value of all samples	/
x_i	measured value of x	/
x'	normalized result	/
$\overline{x}$	average value of x	/
X'	standardized value	/
$\overline{y}$	average value of y	/
y_i	measured value of y	/
Z_t	activation vector of update gate	/
μ	mean of all samples	/
σ	standard deviation of all samples	/
ω_z	weight matrix of the update gate	/
ω_r	weight matrix of the reset gate	/

References

1. Zhu, D.; Yu, X.; Sai, Q.; Wang, S.; Barton, D.; Fieldhouse, J.; Kosarieh, S. Noise and vibration performance of automotive disk brakes with laser-machined M-shaped grooves. *Proc. Inst. Mech. Eng. Part D J. Automob. Eng.* **2022**, *237*, 978–990. [CrossRef]
2. Wang, S.; Guo, W.; Zeng, K.; Zhang, X. Characterization of automotive brake discs with laser-machined surfaces. *Automot. Innov.* **2019**, *2*, 190–200. [CrossRef]
3. Khairnar, H.P.; Phalle, V.M.; Mantha, S.S. Estimation of automotive brake drum-shoe interface friction coefficient under varying conditions of longitudinal forces using Simulink. *Friction* **2015**, *3*, 214–227. [CrossRef]
4. Riva, G.; Varriale, F.; Wahlström, J. A finite element analysis (FEA) approach to simulate the coefficient of friction of a brake system starting from material friction characterization. *Friction* **2021**, *9*, 191–200. [CrossRef]
5. Meng, Y.; Xu, J.; Ma, L.; Jin, Z.; Prakash, B.; Ma, T.; Wang, W. A review of advances in tribology in 2020–2021. *Friction* **2022**, *10*, 1443–1595.
6. Balaji, V.; Lenin, N.; Anand, P.; Rajesh, D.; Raja, V.B.; Palanikumar, K. Brake squeal analysis of disc brake. *Mater. Today Proc.* **2021**, *46*, 3824–3827. [CrossRef]
7. Crolla, D.A.; Lang, A.M. Brake noise and vibration: The state of the art. *Veh. Tribol.* **1991**, *18*, 165–174.
8. Oberst, S.; Lai, J. Chaos in brake squeal noise. *J. Sound Vib.* **2011**, *330*, 955–975. [CrossRef]
9. Jarvis, R.P.; Member, B.M. Vibrations induced by dry friction. *Proc. Inst. Mech. Eng.* **1963**, *178*, 847–857. [CrossRef]
10. Nishiwaki, M. Generalized theory of brake noise. *Proc. Inst. Mech. Eng. Part D J. Automob. Eng.* **1993**, *207*, 195–202. [CrossRef]
11. Zhang, W.; Guo, W.; Liu, X.; Liu, Y.; Zhou, J.; Li, B.; Lu, Q.; Yang, S. LSTM-based analysis of industrial IoT equipment. *IEEE Access* **2018**, *6*, 23551–23560. [CrossRef]
12. Jiang, W.; Zhang, N.; Xue, X.; Xu, Y.; Zhou, J.; Wang, X. Intelligent deep learning method for forecasting the health evolution trend of aero-engine with dispersion entropy-based multi-scale series aggregation and LSTM neural network. *IEEE Access* **2020**, *8*, 34350–34361. [CrossRef]
13. Zhang, X.; Zhang, M.; Xiang, Z.; Mo, J. Research on diagnosis algorithm of mechanical equipment brake friction fault based on MCNN-SVM. *Measurement* **2021**, *186*, 110065. [CrossRef]
14. Zhang, M.; Zhang, X.; Mo, J.; Xiang, Z.; Zheng, P. Brake uneven wear of high-speed train intelligent monitoring using an ensemble model based on multi-sensor feature fusion and deep learning. *Eng. Fail. Anal.* **2022**, *137*, 106219. [CrossRef]
15. Yang, X.; Chen, L. Dynamic state estimation for the advanced brake system of electric vehicles by using deep recurrent neural networks. *IEEE Trans. Ind. Electron.* **2019**, *99*, 9536–9547.
16. Šabanovič, E.; Žuraulis, V.; Prentkovskis, O.; Skrickij, V. Identification of road-surface type using deep neural networks for friction coefficient estimation. *Sensors* **2020**, *20*, 612. [CrossRef] [PubMed]
17. Stender, M.; Tiedemann, M.; Spieler, D.; Schoepflin, D.; Hoffmann, N.; Oberst, S. Deep learning for brake squeal: Brake noise detection, characterization and prediction. *Mech. Syst. Signal Process.* **2021**, *149*, 107181. [CrossRef]
18. Wang, S.; Zhong, L.; Niu, Y.; Liu, S.; Wang, S.; Li, K.; Wang, L.; Barton, D. Prediction of frictional braking noise based on brake dynamometer test and artificial intelligent algorithms. *Proc. Inst. Mech. Eng. Part D J. Automob. Eng.* **2021**, *236*, 2681–2695. [CrossRef]
19. Alexsendric, D.; Barton, D.C. Neural network prediction of disc brake performance. *Tribol. Int.* **2009**, *42*, 1074–1080. [CrossRef]

20. Alexsendric, D.; Barton, D.C.; Vasic, B. Prediction of brake friction materials recovery performance using artificial neural networks. *Tribol. Int.* **2010**, *43*, 2092–2099. [CrossRef]
21. *J2521_201304*; Disc and Drum Brake Dynamometer Squeal Noise Test Procedure. SAE International: Warrendale, PA, USA, 2013.
22. Shalabi, L.A.; Shaaban, Z.; Kasasbeh, B. Data mining: A preprocessing engine. *J. Comput. Sci.* **2006**, *2*, 735–739. [CrossRef]
23. Curtis, A.E.; Smith, T.A.; Ziganshin, B.A.; Elefteriades, J.A. The mystery of the Z-score. *Aorta* **2016**, *4*, 124–130. [CrossRef] [PubMed]
24. Benesty, J.; Chen, J.D.; Huang, Y. On the importance of the Pearson correlation coefficient in noise reduction. *IEEE Trans. Audio Speech Lang. Process.* **2008**, *16*, 757–765. [CrossRef]
25. Reshef, D.N.; Reshef, Y.A.; Finucane, H.K.; Grossman, S.R.; McVean, G.; Turnbaugh, P.J.; Lander, E.S.; Mitzenmacher, M.; Sabeti, P.C. Detecting novel associations in large data sets. *Science* **2011**, *334*, 1518–1524. [CrossRef]
26. Zargar, S. *Introduction to Sequence Learning Models: RNN, LSTM, GRU*; Department of Mechanical and Aerospace Engineering: Colorado Springs, CO, USA, 2021.

lubricants

MDPI

Article

Revealing the Molecular Interaction between CTL Base Oil and Additives and Its Application in the Development of Gasoline Engine Oil

Chunfeng Zhang [1], Xiaojun Zhang [1], Qiang Yan [1], Liyang Wang [1] and Xiangqiong Zeng [2,*]

1 Shanxi Lu'an Taihang Lubrication Technology Co., Ltd., Changzhi 046200, China; zcf224@126.com (C.Z.); zhxjuna3051@yeah.net (X.Z.); 18830063716@163.com (Q.Y.); 15103454013@163.com (L.W.)

2 School of Materials and Chemistry, University of Shanghai for Science and Technology, Shanghai 200093, China

* Correspondence: zengxq202312@163.com

Abstract: In order to improve fuel economy to meet the standard for passenger car oil, a new formulation with good viscosity–temperature performance for gasoline engine oil is required. In this study, coal-to-liquid (CTL) base oil, with a high viscosity index and good low-temperature performance, was selected as the base oil to develop the gasoline engine oil. A systematic study on the molecular interaction between the CTL base oil and the viscosity index improver (VII), including three kinds of hydrogenated styrene diene copolymers (HSD-type) and four kinds of ethylene propylene copolymers (OCP-type), was conducted. It was found that in general, in CTL base oil, the HSD-type VII exhibited a much higher viscosity index, a significantly lower shear stability index, a higher thickening ability, and a lower cold-cranking simulator (CCS) viscosity than that of OCP-type VII. Moreover, when comparing CTL base oil with mineral oil 150N, the combination of CTL base oil and the VII displayed a lower CCS viscosity than that of mineral oil, suggesting it had better low-temperature performance and was able to quickly form a protective oil film on the surface, which was beneficial for the cold start. The functional group distribution state of the VII in base oil was analyzed using synchrotron radiation micro-infrared microscope (SR Micro-IR) technology, which revealed that HSD-1 had a better molecular interaction with CTL6 than 150N because of the better uniformity of the C=C group distribution. Based on this, a SP 0W-20 gasoline engine oil was developed by the combination of CTL base oil and the HSD-1 viscosity index improver, together with an additive package, a polymethacrylate pour point depressant, and a non-silicone defoamer, which showed excellent low-temperature performance, thermal oxidation stability, and detergency performance compared to the reference oil.

Keywords: coal-to-liquid base oil; viscosity index improver; molecular interaction; gasoline engine oil

1. Introduction

With the increasingly stringent emission standards for automobiles and the continuous progress of engine technology, automotive lubricants are required to be developed towards energy conservation, environmental protection, economy, and high-quality levels [1,2]. In particular, the implementation of the passenger car oil standard GF-6B [3] has put forward higher requirements for automotive fuel economy. By using new formulation technologies to reduce viscosity and optimize anti-friction performance of gasoline engine oil, the lubrication and protection of engine oil for the engine can be improved, which is an effective way to improve fuel economy [4,5].

Base oil, as the main component of the lubricant formulation, affects the whole performance and service life of gasoline engine oil, like evaporation loss, pour point, viscosity index, oxidation stability, etc. For instance, due to the requirements of low-viscosity gasoline engine oil, a high-quality base oil with a low viscosity, high viscosity index (VI), and low evaporation loss is preferred [6]. In addition, Kallas M. M. et al. studied the effect of synthetic, semi-synthetic, and mineral base oil of SAE10W40 engine oil on gasoline engine parts wear, and found that synthetic oil showed better performance than semi-synthetic and mineral oil [7].

With the progress in coal catalytic conversion technology and emission reduction technology, coal-to-liquid (CTL) base oil, obtained from coal gasification and then the Fischer–Tropsch catalytic reaction, shows the advantages of a high liquid recovery, low emissions, and cost effectiveness, especially for those countries with coal as the primary raw material [8–11]. Because the structure of CTL base oil is mainly alkanes with a high cetane number and low sulfur, nitrogen, and aromatic content, and the content of isomeric alkanes is much higher than normal alkanes, CTL base oil has a high VI and a low pour point, which makes it a high-quality and clean alternative fuel that can be used to formulate engine oil, compressor oil, hydraulic oil, gear oil, grease, etc. [12,13]. Moreover, CTL can be blended in any ratio with conventional petroleum-based fuels, so it is compatible with the existing automotive technology systems and has promising prospects for synthetic high-grade lube oil [14].

To provide guidance for the application of CTL base oil in the high-grade lube oil market, Yu X. et al. [12] examined the effect of antioxidant additives on the thermal oxidative degradation of CTL base oil and found that diphenylamine (L57) and 2,6-di-tert-butyl-4-methylphenol (T501) can significantly improve the oxidation stability of CTL base oil by inhibiting the formation of carbonyl groups (C=O), with a reduction in the diffusion coefficient of the antioxidant in CTL base oil, the free volume of O_2 in the CTL base oil being reduced, and the antioxidation performance increased. Zhang C. et al. [15] studied the correlation between the molecular structure and viscosity index of CTL base oils. It was found that CTL base oils with a similar carbon number distribution exhibited narrower distillation ranges, lower boiling points, and higher distillation efficiencies than those of mineral base oil, and the average chain length, normal paraffins, and structure S_{67} (6- or 7-methyl-substituted CTL branched structure) caused the CTL base oil to display a higher VI.

Due to the change in structure of CTL base oil compared to traditional base oil, traditional gasoline engine oil additives and/or additive packages will not be applicable to CTL base oil. Therefore, it is very important to explore the molecular interaction between CTL base oil and additives in order to develop CTL-based gasoline engine oil. For reducing viscosity and improving anti-friction performance of gasoline engine oil, various additives including lubrication additives, pour point depressants, viscosity index improvers, etc. were used [16–18]. Among these, the viscosity index improver (VII) is one of the key additives to optimize the viscosity–temperature performance of gasoline engine oil. It indicates the degree of change in viscosity as a function of temperature. A high VI means a small change in viscosity with the change in temperature, so the oil can be thin enough to easily start at low temperatures and remain thick enough to provide good lubrication at high temperatures [19]. Boussaid M. et al. [20] investigated the potential of using polyethylene glycol (PEG) as a VII to improve the VI of paraffinic oil. Polyethylene glycol 1500 was blended with 150NS at a concentration from 0 to 10%. It was found that the optimum performance was obtained for the blend containing 3% PEG, resulting in the highest VI of 114.67 due to good particle distribution, evidenced by a smaller polymer particle size and smaller activation energy difference. It was confirmed that the dispersion of the polymer in the oil was crucial for improving the VI, even in the absence of intermolecular interactions. Mohamad S.A. et al. [21] synthesized six polyacrylate copolymers with C8 to C16 alkyl-chain-length alcohols, and their viscosity-improving performance in base oil SAE 30 was studied under the concentration of 0.5–3%. The results showed that all of the prepared polyacrylate copolymers were effective as VIIs for lube oil, and their effectiveness became more pronounced with the increase in either the concentration, the molecular weight (140,000–236,000), or the alkyl chain length (C8–C12) of the copolymers. Similarly, Ghosh P. et al. [22] compared the VI-enhancing capability of decyl and isooctyl acrylate polymers in two kinds of mineral base oils. The results suggested that the isooctyl acrylate polymer showed a higher VI and pour point depression efficiency than that of the decyl acrylate polymer, which depended on the nature of the mineral base oils as well as the type and concentration of the VIIs. Further, Lomège J. et al. [23] synthesized plant-oil-based amide copolymethacrylate as a VII in an organic triglyceride lube oil. It was found that the thickening power can be improved by increasing the copolymer molecular weight, concentration, dispersity, and pendant aliphatic chain length, or by adding an additional aliphatic chain in the copolymer backbone. In addition, polyvinyl palmitate copolymer has been investigated as a VII in diesel oil [24], MWCNT and ZnO nanoparticles have been studied as VIIs in 5W50 to improve engine oil lubrication in light-duty vehicles [25], a fatty-acid-based comb (co) poly(9-alkyl 12-hydroxystearate) was synthesized by click chemistry and studied as a VII in a mineral-based lubricant [26], and a core cross-linked star copolymer was synthesized and evaluated as a VII for application in hydraulic lubricants [27]. In addition, Khalafvandi et al. [28] studied the viscosity properties of three types of VIIs (ethylene propylene, star isoprene, and two di-block styrene isoprene) in different types of base oils (API Groups I, II, and III). It was found that intrinsic viscosity depended on the polymer molecular weight, and the size of the polymer depended on the solubility of the polymer in the base oil. Although various VIIs have been developed and their performances in different base oils have been evaluated, their suitability for use in CTL base oil has not been revealed yet.

Therefore, the aim of this study was to develop a kind of gasoline engine oil using new CTL base oil, based on the exploration of the interaction between the CTL base oil and one of the key additives, the VII. The interaction between the CTL base oil and two types of VIIs was investigated systematically, including three kinds of hydrogenated styrene diene copolymers (HSD-type) and four kinds of ethylene propylene copolymers (OCP-type). The functional group distribution state of the viscosity index improver in the CTL base oil was revealed using synchrotron radiation micro-infrared microscope technology. Based on this, a SP 0W-20 gasoline engine oil was developed, and its low-temperature performance, thermal oxidation stability, and detergency performance were evaluated. The reason for choosing SP 0W-20 as our target was its high market share and moreover, new engines are designed to use lower-viscosity oils to reduce friction losses, since it was found that lowering lube oil viscosity from 10W40 to 0W10 yields a 4% improvement in fuel economy [29].

2. Materials and Methods

2.1. Materials

The base oils used in this study included coal-to-liquid base oils (CTL4 and CTL6, both prepared through coal gasification and then the Fischer–Tropsch catalytic reaction by Shanxi Lu'an company, and the distribution of carbon number of CTL4 and CTL6 is shown in Table S1), Ketjenlube 15 (KL15, dibasic acid ester, supplied by Italmatch Chemicals), and 150N (supplied by Formosa Plastics Group, and the distribution of the carbon number of 150N is shown in Table S1 as well). Their physical and chemical properties are displayed in Table 1, and the test equipment and method for each test are listed in Table S2. The additives used included an additive package (AP, supplied by Afton) with dark brown color, density of 0.935 g/mL at 15 °C, flash point of at least 175 °C, kinematic viscosity of 173 mm^2/s at 100 °C, total base number of 62 mg KOH/g, and element contents of P (0.57 wt.%), Ca (0.99 wt.%), Zn (0.65 wt.%), N (0.96 wt.%), Mg (0.33 wt.%), and Mo (0.03 wt.%), a polymethacrylate pour point depressant (PPD) with a kinematic viscosity of 189.6 mm^2/s at 100 °C, and a non-silicone defoamer (DF) with a kinematic viscosity of 8.548 mm^2/s at 100 °C. The viscosity index improvers we studied included three kinds of hydrogenated styrene diene copolymers (HSD-1, HSD-2, and HSD-3) and four kinds of ethylene propylene copolymers (OCP-1, OCP-2, OCP-3, and OCP-4). Their code and supplier information is listed in Table S3. Their physical and chemical properties were evaluated and analyzed, as shown in Sections 2.3 and 3.1.

Table 1. The physical and chemical properties of the base oils used.

Name	Kinematic Viscosity (100 °C, mm²/s)	Kinematic Viscosity (40 °C, mm²/s)	Viscosity Index	CCS Viscosity (−30 °C, mPa·s)	Pour Point (°C)	Flash Point by Open Cup (°C)	Evaporation Loss (NOACK, 250 °C, 1 h, %)
CTL4	4.155	17.80	141	922	−33	210	11.3
CTL6	6.181	31.34	150	2497	−36	247	3.9
KL15	5.350	27.00	138	4151	−63	235	7.5
150N	5.258	29.17	112	4483	−24	229	12.4

2.2. *Preparation of the Base Oil and Viscosity Index Improver Blends*

To study the interaction between viscosity index improver and the base oil, the base oil was blended with the viscosity index improver to make sols, and their solubility and stability, thickening ability, various kinematic viscosity, cold-cranking simulator viscosity index, and shear stability index were evaluated and analyzed.

The blends were made as follows: Firstly, 88 g of base oil was placed in a beaker and heated to 85 ± 5 °C. Then, under stir, 12 g viscosity index improver (VII solid) was slowly added to the base oil (in a ratio of 12 wt.% VII: 88 wt.% base oil) and heated to 135 °C. The solution was stirred at 135 °C for at least 8 h until the viscosity index improver dissolved completely. Stirring was stopped and the solution was maintained at 110 °C for another 8 h, then cooled down to obtain the liquid samples (VII liquid). The liquid sample was mixed with the base oil in a ratio of 1:9 (VII liquid : base oil) to obtain the sol samples (VII sol). The base oil used included CTL6 and 150N, respectively.

2.3. *The Characterization of the Viscosity Index Improver Blends*

2.3.1. The Solubility Test

The solubility of the VIIs in the base oils were tested by observing the appearance of the blends (VII sols) after storing at different temperatures (0 °C, −10 °C, −20 °C, −30 °C, and −40 °C) for 2 h, one after the other, and then recovering back to room temperature (rt.).

2.3.2. The Physico-Chemical Property Measurements

The physico-chemical properties of the VII blends (VII sols) were measured according to standard methods, including the kinematic viscosity (40 °C, 100 °C, mm²/s), VI, cold-cranking simulator viscosity (CCS, −20 °C, mPa·s), kinematic viscosity after shearing (100 °C, mm²/s), thickening ability (mm²/s), and shear stability index (SSI). The test equipment and method for each test are listed in Table S2. The thickening ability was calculated based on the kinematic viscosity at 100 °C according to SH/T 0622 with the following equation.

$$T = \nu_t - \nu_b$$

Here, T is the thickening ability (mm²/s), ν_t is the kinematic viscosity of the base oil with the viscosity index improver at 100 °C (mm²/s), and ν_b is the kinematic viscosity of the base oil at 100 °C (mm²/s).

In addition, the one-way ANOVA test and multiple t-test were used to conduct the statistical analysis. In the figures in Section 3, one asterisk (*) indicates significance at $p < 0.05$, and two asterisks (**) indicate significance at $p < 0.01$.

2.3.3. The Synchrotron Radiation Micro-Infrared Test

Synchrotron radiation micro-infrared test is a useful technique to illustrate the chemical group distribution of molecules. In this study, we applied the synchrotron radiation micro-infrared (SR Micro-IR) at BL01B beamline in Shanghai Synchrotron Radiation Facility (SSRF) to analyze the chemical group distribution of the VII in the base oil.

2.4. Preparation of the Gasoline Engine Oil

The SP 0W-20 gasoline engine oil developed in this work was prepared as follows: The base oil (CTL4, CTL6, KL15) was added to the reaction kettle, then stirred and heated to 60 °C. Then, the additive VII liquid (12 wt.% VII mixed with 88 wt.% CTL6, as described in Section 2.2), AP, PPD, and DF was added to the base oil in turn. The temperature was maintained at 60 °C while stirring for 2 h, then cooled down while stirring to obtain the SP 0W-20 gasoline engine oil samples.

2.5. The Physico-Chemical Property Measurements of the Developed Gasoline Engine Oil

The physico-chemical properties of the developed gasoline engine oil were measured according to standard methods, including the kinematic viscosity (100 °C, mm^2/s), VI, low-temperature dynamic viscosity (−35 °C, mPa·s), low-temperature pumping viscosity (−40 °C, mPa·s), high temperature and high shear viscosity (150 °C, mPa·s), evaporation loss (%) and foam property (ml/mL), pour point, flash point, etc. The test equipment and method for each test are listed in Table S2.

2.6. The Performance Evaluation of the Developed Gasoline Engine Oil

2.6.1. The Thermal Oxidation Stability Test

The thermal oxidation stability of the developed gasoline engine oil was tested by Crankcase Simulation test to simulate the lacquer and coking formation of the oil during piston working, according to SH/T 0300, similar to FTM 791-3462, using internal combustion engine oil coking tendency tester C-9 supplied by Shanghai Rundi Scientific Instrument Co., Ltd. (Shanghai, China). The tested oil sample was sprayed onto an aluminum sheet at a high temperature, and due to the oxidation of the oil, a coking film was formed on the sheet. During the test, the oil samples and aluminum plate were maintained at 150 °C and 320 °C, respectively, in the Crankcase simulation tester, and operated for 6 h. After this, the color and scale level of the lacquer and coking film were assessed on a 1-to-10 scale according to the reference card, and the coking film weight on the aluminum plate was evaluated, which was correlated to the thermal oxidation stability of the oil [30].

2.6.2. The Detergency Test

The detergency of the developed gasoline engine oil was tested by the hot tube method according to SH/T 0645, using a hot tube tester R1091G supplied by Shanghai Rundi Scientific Instrument Co., Ltd. The oil sample and oxygen were mixed together and circulation refluxed in hot tube at 275 °C for 4 h. Then, the deposits generated in the glass tube were assessed by ranking the color and length of the formed lacquer on a scale of 0 to 10 according to the reference card, which was correlated to the detergency of the test oil [30].

3. Results and Discussion

3.1. The Interaction between the Viscosity Index Improver and the Base Oil

3.1.1. The Solubility Analysis

The VII and the base oil were blended together to make sols as described in Section 2.2. The appearance of the blends (VII sols) under different temperatures was evaluated and is exhibited in Table 2, in which the base oils used for the stability tests were CTL6 and 150N, respectively. From the images in Table 2, we can see that all the VIIs could be dispersed well in both base oils, with a transparent white to light yellow appearance after being stored at 0 °C and −10 °C for 2 h. When the storing temperature was reduced to −20 °C, the sols became translucent, while when the temperature increased to room temperature again, the sols turned transparent again, indicating the good solubility of the VIIs in both base oils.

Table 2. The appearance of the blends under different temperatures.

Base Oil	Temperature (°C)	HSD-1	HSD-2	HSD-3	OCP-1	OCP-2	OCP-3	OCP-4
CTL6	0							
	−10							
	−20							
	rt.							
150N	0							
	−10							
	−20							
	rt.							

3.1.2. The Influence of VII Concentration on the Viscosity of the Blends

The VII is used to optimize the viscosity–temperature performance. As demonstrated by Coutinho [19], the viscosity of a VII polymer solution depends on many factors, including the chain size of the VII polymer, the concentration, the nature of the base oil, the strength of the inter- and intra-molecular interactions among the VII polymer molecules, and the degree of the interaction between the VII polymer molecules and the base oil. Therefore, firstly we investigated the effect of the VII concentration on the viscosity of the blends (VII liquid). Here, the VII liquid was made by mixing the VII solid (HSD-1 and OCP-1, respectively) with the base oil (CTL6 and 150N, respectively) at concentrations of 0.5%, 1.0%, 1.5%, 2.0%, and 3.0% using the same method as described in Section 2.2.

The results are illustrated in Figure 1. It can be found that no matter which VII and base oils used, with the increase in VII concentration, the kinematic viscosity at both 40 °C (Figure 1a) and 100 °C (Figure 1b), and the viscosity index (Figure 1c) were all increased.

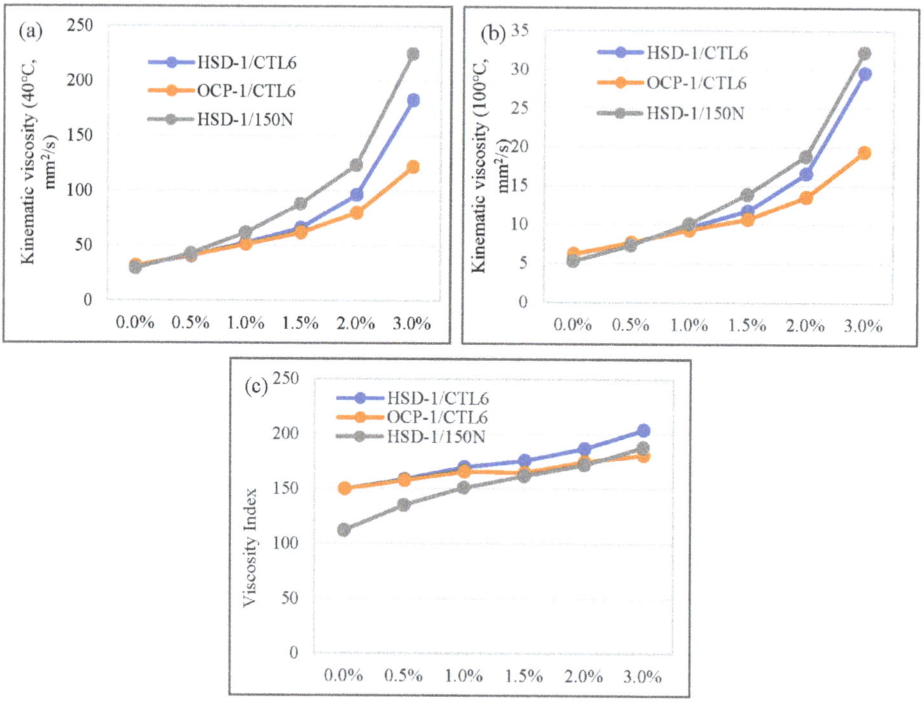

Figure 1. The influence of VII concentration on the viscosity of the blends: (**a**) Kinematic viscosity at 40 °C; (**b**) Kinematic viscosity at 100 °C; (**c**) Viscosity index.

When comparing the HSD-type VII (HSD-1) with the OCP-type VII (OCP-1) in CTL6, it can be seen that the kinematic viscosity of HSD-1 was higher than that of OCP-1 at both 40 °C and 100 °C, and the viscosity index of HSD-1 was higher than that of OCP-1, indicating HSD-1 had a smaller change in viscosity with the change in temperature than that of OCP-1 in the CTL6 base oil.

When comparing HSD-1 in the CTL6 and 150N base oils, we found that the kinematic viscosity of HSD-1 in 150N was higher than that in the CTL6 base oil at both 40 °C and 100 °C, while the viscosity index of HSD-1 in 150N was smaller than that in the CTL6 base oil, indicating that when used in the 150N base oil, the HSD-1 has a higher change in viscosity with the change in temperature.

Therefore, it can be concluded that HSD-1 had better viscosity–temperature performance than that of OCP-1, and it had better performance in CTL6 than in 150N. Usually, as the temperature increased the extent of the intra- and intermolecular interactions of VII and base oil reduced [19]. The results suggested that CTL6 was a good solvent for the HSD-1-type VII due to its 'pure' hydrogenated polyisoprene block, while 150N was a poor solvent for the HSD-1-type VII. In the good solvent CTL6, HSD-1 (hydrogenated styrene diene copolymer) was well dispersed, with the polymer chains uncoiled. Therefore, as the temperature rose, the reduction in bond strength was consequently low, so that the viscosity index was high. In the poor solvent 150N, the temperature increase may have provoked the increase in hydrodynamic volume of the HSD-1 polymer molecules and hence an increase in viscosity and a decrease in viscosity index compared to that in CTL6 [19].

3.1.3. The Influence of Type of VII on the Shear Stability of the Blends

In an automative engine, lubricants are subjected to mechanical shearing like engine running. If the shearing force is too strong to break some chemical bonds in the VII polymer, then its molecular weight will reduce, resulting in a loss in its efficiency, affecting the quality level and service life of gasoline engine oils [19]. Therefore, shear stability is an important property of a VII. According to the API standard, the shear stability index of gasoline engine oil should be no more than 35 for GF-6 vehicles.

Firstly, the kinematic viscosity of the blends (VII sols) before and after shearing was measured and is displayed in Figure 2a,b. The kinematic viscosity of all samples was reduced after shearing, and in general, the viscosity-reducing degree of the OCP-type VII was higher than that of the HSD-type VII. Also, this phenomenon was further confirmed by the shear stability index (SSI), as illustrated in Figure 2c. In addition, the statistical analysis indicated that there was a significant difference among different types of VIIs. For the kinematic viscosity at 100 °C before shearing, between HSD-1 and OCP-3, OCP-1 and OCP-2, and OCP-1 and OCP-4, there was no obvious difference in either CTL6 and 150N, HSD-2 and HSD-3 showed no obvious differences in CTL6, and OCP-2 and OCP-4 showed no obvious differences in 150N. For the kinematic viscosity at 100 °C after shearing, OCP-1 and OCP-2 showed no obvious differences in either CTL6 or 150N; additionally, OCP-1 and OCP-3, and OCP-2 and OCP-3 showed no obvious differences in 150N, and HSD-2 and HSD-3, OCP-1 and OCP-4, and OCP-2 and OCP-4 showed no obvious differences in CTL6. The results suggested that the behavior of OCP-1 from Arlanxeo and OCP-2 from Shenzhen Kunvii Petrochemical Technology were almost similar with respect to kinematic viscosity at 100 °C before and after shearing in both CTL6 and 150N. The rest of the samples all had significant differences.

From Figure 1c, it can be seen that the SSI of the VII in 150N was higher than that in CTL6; there was an especially significant difference between 150N and CTL6 for OCP-1 ($p < 0.05$), OCP-3 ($p < 0.01$), and OCP-4 ($p < 0.05$), indicating that the CTL and VII blends showed better shear stability than the mineral oil and VII blends. Moreover, the statistical analysis found that the SSIs of all the HSD-type VIIs were significantly lower than those of all the OCP-type VIIs in both CTL6 and 150N. Especially, HSD-1 showed the lowest SSI of 2.30 in the CTL6 base oil and 2.95 in the 150N base oil, suggesting the better shear stability of the HSD-1-type VII in CTL6. Within the groups, HSD-2 and HSD-3, and OCP-1 and OCP-2 showed no obvious differences in either CTL6 or 150N, and OCP-4 showed no obvious differences between OCP-1 and OCP-2 in 150N. Therefore, it can be concluded that the HSD-type VII and CTL base oil blends were promising in the development of gasoline engine oil in terms of shear stability.

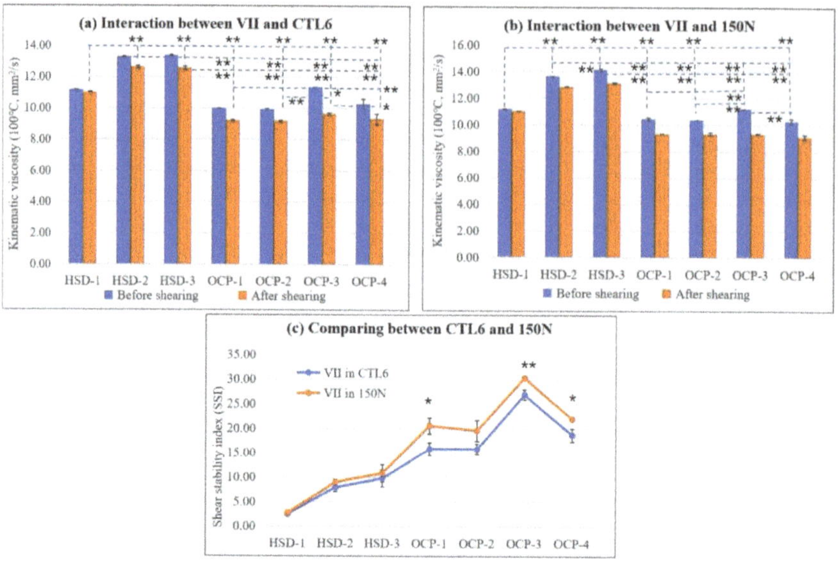

Figure 2. The shear stability of the VII and base oil blends: (**a**) Kinematic viscosity at 100 °C before and after shearing for VII in CTL6; (**b**) Kinematic viscosity at 100 °C before and after shearing for VII in 150N; (**c**) Shear stability index. (one asterisk (*) indicates significance at $p < 0.05$; two asterisks (**) indicate significance at $p < 0.01$).

3.1.4. The Influence of Type of VII on the Thickening Ability of the Blends

Thickening ability is another important parameter to evaluate the performance of a VII. Figure 3 exhibits the thickening ability of different types of VIIs in both CTL6 and 150N (VII sols). It can be seen that in general, the thickening ability of the VII in 150N was higher than that in CTL6, and there was significant a difference in thickening ability between 150N and CTL6 for HSD-2, HSD-3, OCP-1, and OCP-2.

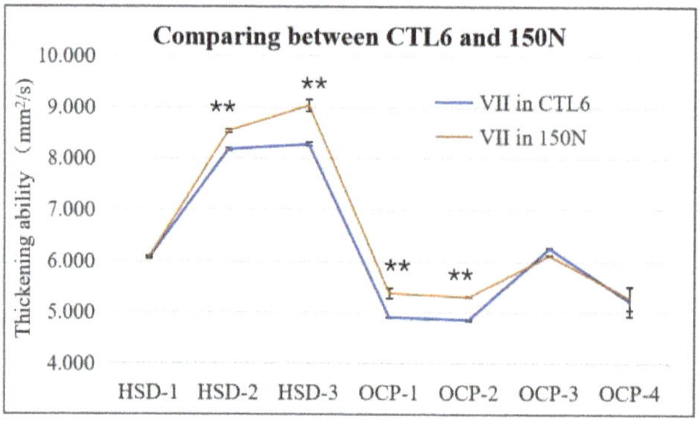

Figure 3. The thickening ability of the VII and base oil blends (two asterisks (**) indicate significance at $p < 0.01$).

In addition, no matter in which base oil, the HSD-type VII showed higher thickening ability than the OCP-type VII, and significant difference could be found between the HSD-type VII and the OCP-type VII, except for HSD-1 and OCP-3. Within the groups, there was no obvious difference between HSD-2 and HSD-3 in CTL6, between OCP-1 and OCP-2 in either CTL6 or 150N, between OCP-1 and OCP-4 in 150N, or between OCP-2 and OCP-4 in

150N. The rest of the samples all had significant differences. The better thickening ability of the HSD-type VII than the OCP-type VII may be due to the different networks formed. As a hydrogenated styrene diene copolymer, the HSD-type VII may form a polystyrene-block-associated network, while as an ethylene propylene copolymer, the OCP-type VII may form a moderately expanded dispersed network [19].

3.1.5. The Influence of Type of VII on the Low-Temperature Performance of the Blends

A VII can improve the viscosity–temperature performance of oil, but in the meantime, it may increase the low-temperature viscosity, affecting the cold start performance of the oil. Therefore, it is important to study the influence of VIIs on low-temperature performance.

Cold-cranking simulator viscosity (CCS, −20 °C, mPa·s) is a parameter showing the torque necessary to turn the crankshaft in the lubricant, which is correlated to the cold start property. As can be seen in Figure 4 for the VII sols, no matter in which base oil, the CCS of the HSD-type VII was significantly lower than that of the OCP-type VII, except between HSD-1, HSD-3, and OCP-4. Moreover, no matter which VII was used, the CCS in the CTL6 base oil was lower than that in 150N, and there was a significant difference in the CCS between 150N and CTL6 for HSD-2, OCP-1, OCP-2, and OCP-3, suggesting that the CTL base oil with the addition of the HSD-type VII had the better low-temperature performance, with a better low-temperature fluidity beneficial for a cold start.

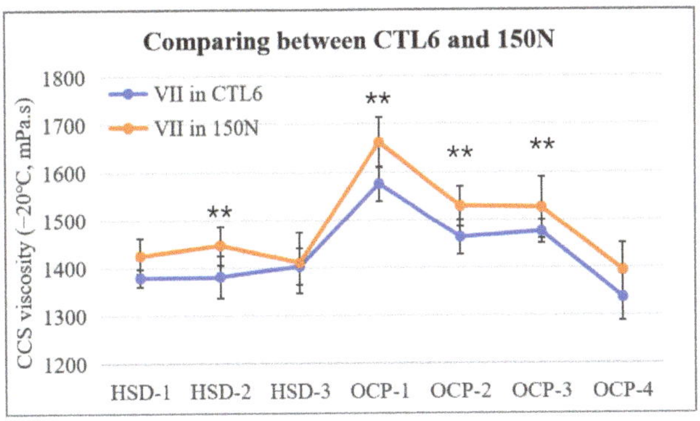

Figure 4. The low-temperature performance of the VII and base oil blends (two asterisks (**) indicate significance at $p < 0.01$).

3.2. *The Characterization of the Molecular Interaction between the Viscosity Index Improver and the Base Oil*

To reveal the molecular interaction between the VII and base oil, Pirouz et al. [31] applied pyrene excimer fluorescence to quantitatively measure the molar fraction of the intermolecular interactions (f_{inter}) between ethylene propylene copolymers in toluene in the presence of wax, and Gholami et al. [32] used fluorescently labeled PPD to probe the interactions between PPD, VII, and wax in octane. In this study, the chemical group distribution of VII in base oil (VII sols) was analyzed using the synchrotron radiation micro-infrared (SR Micro-IR) technique.

The OCP-type VII is an ethylene propylene copolymer; therefore, it has the IR characteristic absorption peaks of the -CH_2- chemical group at around 2915 cm^{-1} for the asymmetric stretching vibration, around 2846 cm^{-1} for the symmetric stretching vibration, and around 1459 cm^{-1} for the bending vibration. However, due to the similar IR characteristic absorption peaks of the OCP-type VII in both CTL and mineral oil, it is hard to analyze the distribution of the chemical group of the OCP-type VII in base oil. Meanwhile, the HSD-type VII is a hydrogenated styrene diene copolymer, which has the IR characteristic absorption peaks of the benzene ring skeleton (C=C) stretching vibration at around 1600 cm^{-1} and 1500 cm^{-1}. Therefore, it is valuable to analyze the chemical group distribution of the HSD-type VII in base oils. Also, the test results of the C=C stretching vibration absorption peak (around 1490 cm^{-1}) distribution of the HSD-type VII in base oil are shown in Figure 5. It is obvious that in CLT6, the uniformity of the C=C group distribution was ranked as HSD-1 ≈ HSD-2 > HSD-3, which was consistent with the performance of the HSD and base oil blends. In addition, the uniformity of the C=C group distribution in CTL6 (Figure 5(a2)) was much better than that in 150N (Figure 5(d2)), indicating a better molecular interaction for HSD-1/CTL6 than that for HSD-1/150N [33,34].

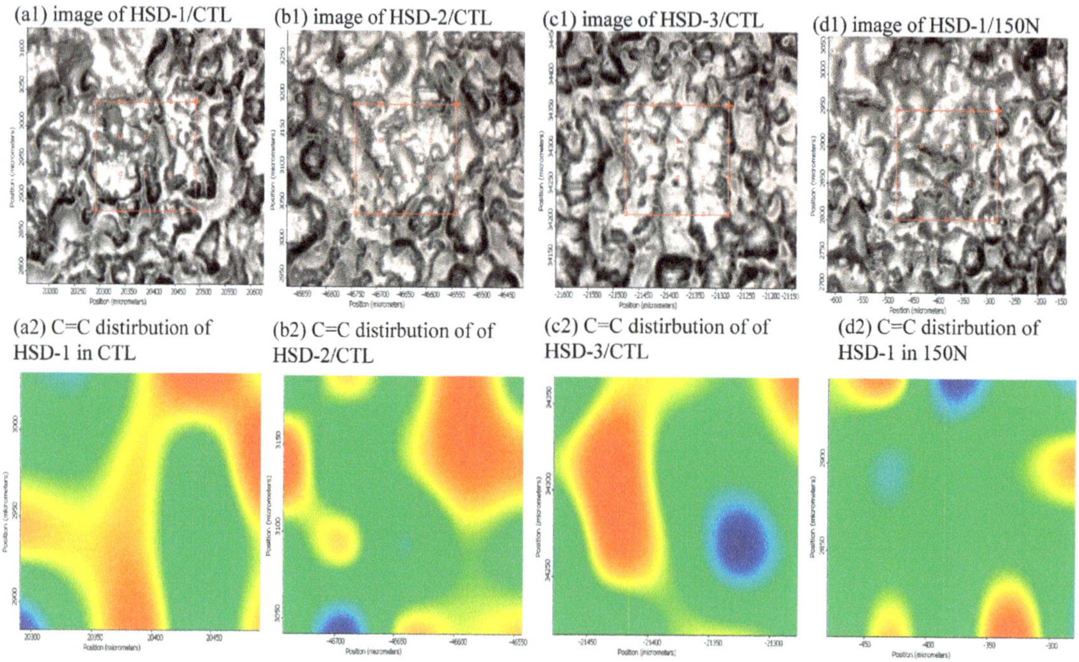

Figure 5. The chemical group distribution of HSD−type VII in base oils.

3.3. *The Influence of the Concentration of the Viscosity Index Improver on the Physico-Chemical Properties of the Gasoline Engine Oil*

Based on the study on the interaction between the VII and base oil, the HSD-1 VII was selected to develop the gasoline base oil. The formulation of the developed gasoline engine oil is displayed in Table 3, in which the concentration of the VII liquid and the ratio of the base oils were varied.

Table 3. The formulation of the developed gasoline engine oil.

Items	CTL4	CTL6	KL15	HSD-1 Liquid	AP	PPD	DF	In Total
F#1	62.59%	10.00%	10.00%	4.00%	13.10%	0.30%	0.01%	100.00%
F#2	64.59%	10.00%	10.00%	2.00%	13.10%	0.30%	0.01%	100.00%
F#3	12.00%	62.59%	10.00%	2.00%	13.10%	0.30%	0.01%	100.00%

Their physico-chemical properties compared to reference oil SP 0W-20 are exhibited in Table 4. Firstly, we compared F#2 and F#3 with the difference between the CTL4 and CTL6 base oil ratio, and it can be seen that although all the properties were within the qualification, the low-temperature dynamic viscosity of F#3 with a higher CTL6 ratio was

6100 mPa·s, approaching the upper limit of 6200 mPa·s, and was much higher than that of the reference SP 0W-20 oil (3381 mPa·s), indicating a risk to low-temperature performance. Therefore, we increased the ratio of the CTL4 base oil and compared the influence of VII concentration, using 2.00% VII for F#2 and 4.00% VII for F#1. It was found that with the increase in CTL4 ratio, the kinematic viscosity of the formulation reduced dramatically from 8.45 mm^2/s to 6.60 mm^2/s, and the high temperature and high shear viscosity reduced from 2.817 mPa·s to 2.519 mPa·s, even lower than the qualification. To solve this problem, we increased the concentration of the VII (F#1). Also, from the physico-chemical properties shown in Table 4, we can see that the viscosity characteristics at different conditions of F#1were comparable to that of the reference oil SP 0W-20. Moreover, the pour point, flash point, evaporation loss, and foam properties of the developed gasoline engine oil F#1 were even better than that of the reference oil SP 0W-20, indicating a better low-temperature fluidity and cold start property, safety, and low oil consumption characteristics. Therefore, F#1 was chosen to be the final gasoline engine oil formulation.

Table 4. The physico-chemical properties of the developed gasoline engine oil.

Items	Qualification	F#1	F#2	F#3	Reference SP 0W-20	Test Method
Kinematic viscosity (100 °C, mm^2/s)	5.6–9.3	8.90	6.60	8.45	8.76	GB/T 265 [35]
Viscosity index	report	177	168	170	171	GB/T 1995 [36]
Pour point, °C	≤−40	−48	−45	−45	−51	GB/T 3535 [37]
Flash point (opening), °C	≥200	235	231	228	229	GB/T 3536 [38]
Low-temperature dynamic viscosity (−35 °C, mPa·s)	≤6200	4636	4530	6100	3381	GB/T 6538 [39]
Low-temperature pumping viscosity (−40 °C, mPa·s, no yield stress)	≤60,000	14,887	11,562	22,624	16,300	SH/T 0562 [40]
High temperature and high shear viscosity (150 °C, mPa·s)	≥2.6	2.800	2.519	2.817	2.611	SH/T 0703 [41]
Evaporation loss (%)	≤15	8.30	8.42	5.80	10.0	SH/T 0059 [42]
Foam property (ml/mL) Procedure I (24 °C) Procedure II (93.5 °C) Procedure III (Last 24 °C)	 ≤10/0 ≤50/0 ≤10/0	 0/0 10/0 0/0	 0/0 10/0 0/0	 0/0 15/0 0/0	 0/0 20/0 0/0	GB/T 12579 [43]

3.4. *The Performance of the Developed Gasoline Engine Oil*

During the use of gasoline engine oil, certain components with poor thermal stability in the oil will gradually be oxidized due to long-term exposure to high-temperature and high-shear environments. As the degree of oxidation becomes severe, the oxidation products will further react and condense, ultimately generating macro-molecular substances that are insoluble in oil, known as high-temperature deposits, which will significantly affect the performance of the gasoline engine oil. Therefore, the thermal oxidation stability and detergency performance related to the high-temperature deposits were evaluated for the developed gasoline engine oil F#1.

3.4.1. The Thermal Oxidation Stability

The thermal oxidation stability of the developed gasoline engine oil was tested using Crankcase Simulation test, and the results are shown in Table 5. It was obvious that the developed gasoline engine oil F#1 had better thermal oxidation stability than that of the reference oil SP 0W-20, with smaller amounts of black deposits, with a lighter and more even color.

Table 5. The thermal oxidation stability of the developed gasoline engine oil.

Items	F#1	Reference SP 0W-20
Deposit weight (mg)	5	15
Color scale	1 + 1	3 + 0
Results description	Even, light yellow lacquer, carbon deposits level 1	Uneven bright patterns, carbon deposits level 2
Images of the plate surface		

3.4.2. The Detergency Performance

The detergency performance of the developed gasoline engine oil was tested using the hot tube test and the results are illustrated in Table 6. After the tests, both the developed gasoline engine oil and the reference oil SP 0W-20 showed a similar yellow lacquer with the color scale being the same at 6.5, suggesting the developed engine oil had similar detergency performance to that of the reference oil.

Table 6. The detergency performance of the developed gasoline engine oil.

Items	F#1	Reference SP 0W-20
Results description	Yellow lacquer	Yellow lacquer
Color scale	6.5	6.5
Images of the tube		

4. Conclusions

A gasoline engine oil SP 0W-20 was developed using CTL base oil and the HSD-type VII based on studying the interaction between the VII and base oil. The following conclusions can be drawn.

1. The interaction between three kinds of HSD-type VIIs, four kinds of OCP-type VIIs, and CTL and mineral base oils were systematically evaluated. It was found that firstly, all samples showed good solubility after storing at different temperatures. Secondly, no matter which VII and base oils were used, the kinematic viscosity and the viscosity index all increased with the increase in VII concentration. Also, the HSD-type VII illustrated a higher VI, smaller SSI, better thickening ability, and lower CCS viscosity than that of the OCP-type VII, and the HSD-type VII and CTL base oil combination exhibited the best shearing stability and low-temperature performance, which is promising for the development of gasoline engine oil.
2. The SR Micro-IR analysis revealed that HSD-1 had a better molecular interaction with CTL6 than 150N because of better uniformity of the C=C group distribution.
3. By adjusting the base oil ratio and VII concentration, a gasoline engine oil F#1 was developed, with comparable viscosity characteristics and detergency performance to that of the reference oil SP 0W-20, and with better low-temperature fluidity and cold start property, lower oil consumption characteristic and better thermal stability than that of the reference oil SP 0W-20.

Supplementary Materials: The following supporting information can be downloaded at: https://www.mdpi.com/article/10.3390/lubricants12080275/s1, Table S1: The chemical composition of the CTL base oils. Table S2: The equipment and test method information of all tests in the study. Table S3: The code and supplier information of VIIs used in the study.

Author Contributions: Methodology, X.Z. (Xiaojun Zhang); Validation, L.W.; Investigation, Q.Y.; Writing—original draft, C.Z.; Supervision, X.Z. (Xiangqiong Zeng). All authors have read and agreed to the published version of the manuscript.

Funding: This research received no external funding.

Data Availability Statement: The original contributions presented in the study are included in the article, further inquiries can be directed to the corresponding author.

Acknowledgments: We thank the BL01B beamline of the National Facility for Protein Science in Shanghai (NFPS) at Shanghai Synchrotron Radiation Facility, for the support in the synchrotron infrared micro-spectroscopy measurements (No. 2022-NFPS-PT-007342).

Conflicts of Interest: Authors Chunfeng Zhang, Xiaojun Zhang, Qiang Yan and Liyang Wang were employed by the company Shanxi Lu'an Taihang Lubrication Technology Co., Ltd. The remaining author declares that the research was conducted in the absence of any commercial or financial relationships that could be construed as a potential conflict of interest.

References

1. Tóth, Á.D.; Szabó, Á.I.; Leskó, M.Z.; Rohde-Brandenburger, J.; Kuti, R. Tribological properties of the nanoscale spherical Y_2O_3 particles as lubricant additives in automotive application. *Lubricants* **2022**, *10*, 28. [CrossRef]
2. Arumugam, S.; Ellappan, R.; Sriram, G. Degradation of engine components upon exposure to chemically modified vegetable oil-based automotive lubricant. *J. Indian Chem. Soc.* **2021**, *98*, 100227. [CrossRef]
3. *GF-6B:API 1509*; Engine Oil Licensing and Certification System. API Publishing Services: Washington, DC, USA, 2019.
4. Srivastava, G. Framework for brand positioning of automotive lubricants by using structural equation modelling. *Int. J. Manag. Pract.* **2023**, *16*, 89–103. [CrossRef]
5. Tóth1, Á.D.; Knaup, J. Investigation of the tribological properties of nano-scaled ZrO_2 and CuO additive in automotive lubricants. *IOP Conf. Ser. Mater. Sci. Eng.* **2020**, *903*, 012015. [CrossRef]
6. Kang, G.; Zhang, F.; Liu, H.; Wei, B.; Jin, Z.; Lv, H. Rapid separation and API grades identification of base oil in low viscosity gasoline engine oil SN 0W-16. *ACS Omega* **2024**, *9*, 21270–21275. [CrossRef]
7. Kallas, M.M.; Al Sabek, M.S.; Saoud, Y. Experimental comparison of the effect of using synthetic, semi-synthetic, and mineral engine oil on gasoline engine parts wear. *Adv. Tribol.* **2024**, *2024*, 5997292. [CrossRef]

8. Tong, R.; Zhang, B.; Yang, X.; Wang, Y.; Zhang, L. A life cycle analysis comparing coal liquefaction techniques: A health-based assessment in China. *Sustain. Energy Technol. Assess.* **2021**, *44*, 101000. [CrossRef]
9. Zhang, Y.; Li, J.; Yang, X. Comprehensive competitiveness assessment of four coal-to-liquid routes and conventional oil refining route in China. *Energy* **2021**, *235*, 121442. [CrossRef]
10. Song, M.; Wang, Z.; Wang, W. An empirical study on technological innovation and corporate competitiveness of listed coal-to-liquids companies in China. *Front. Environ. Sci.* **2022**, *10*, 1043094. [CrossRef]
11. Kong, Z.; Dong, X.; Jiang, Q. Forecasting the development of China's coal-to-liquid industry under security, economic and environmental constraints. *Energy Econ.* **2019**, *80*, 253–266. [CrossRef]
12. Yu, X.; Zhang, C.; Wang, H.; Wang, W.; Jiang, C.; Peng, C.; Yang, K. Oxidation degradation analysis of antioxidant added to CTL base oils: Experiments and simulations. *J. Therm. Anal. Calorim.* **2023**, *148*, 7033–7046. [CrossRef]
13. Zhang, Z.; Zhang, C.; Cai, P.; Jing, Z.; Wen, J.; Li, Y.; Wang, H.; An, L.; Zhang, J. The potential of coal-to-liquid as an alternative fuel for diesel engines: A review. *J. Energy Inst.* **2023**, *109*, 101306. [CrossRef]
14. Sun, W.; Sun, Y.; Guo, L.; Zhang, H.; Yan, Y.; Zeng, W.; Lin, S. Comparative assessment of n-butanol addition in CTL on performance and exhaust emissions of a CI engine. *Fuel* **2021**, *303*, 121223. [CrossRef]
15. Zhang, C.; Wang, H.; Yu, X.; Peng, C.; Zhang, A.; Liang, X.; Yan, Y. Correlation between the molecular structure and viscosity index of CTL base oils based on ridge regression. *ACS Omega* **2022**, *7*, 18887–18896. [CrossRef]
16. Markandan, K.; Nagarajan, T.; Walvekar, R.; Chaudhary, V.; Khalid, M. Enhanced tribological behaviour of hybrid $MoS_2@Ti_3C_2$ MXene as an effective anti-Friction additive in gasoline engine oil. *Lubricants* **2023**, *11*, 47. [CrossRef]
17. Rahimi, B.; Semnani, A.; Nezamzadeh-Ejhieh, A.; Langeroodi, H.S.; Davood, M.H. Monitoring of the physical and chemical properties of a gasoline engine oil during its usage. *J. Anal. Methods Chem.* **2012**, *2012*, 819524. [CrossRef]
18. Kaleli, E.H.; Demirtas, S. Experimental investigation of the effect of tribological performance of reduced graphene oxide additive added into engine oil on gasoline engine wear. *Lubr. Sci.* **2023**, *35*, 118–143. [CrossRef]
19. Coutinho, F.M.B.; Teixeira, S.C.S. Polymers used as viscosity index improvers A comparative study. *Polym. Test.* **1993**, *12*, 415–422. [CrossRef]
20. Boussaid, M.; Haddadine, N.; Benmounah, A.; Dahal, J.; Bouslah, N.; Benaboura, A.; El-Shall, S. Viscosity-boosting effects of polymer additives in automotive lubricants. *Polym. Bull.* **2023**, *81*, 6995–7011. [CrossRef]
21. Mohamad, S.A.; Ahmed, N.S.; Hassanein, S.M.; Rashad, A.M. Investigation of polyacrylates copolymers as lube oil viscosity index improvers. *J. Pet. Sci. Eng.* **2012**, *100*, 173–177. [CrossRef]
22. Ghosh, P.; Das, M. Study of the influence of some polymeric additives as viscosity index improvers and pour point depressants-synthesis and characterization. *J. Pet. Sci. Eng.* **2014**, *119*, 79–84. [CrossRef]
23. Lomège, J.; Mohring, V.; Lapinte, V.; Negrel, C.; Robin, J.J.; Caillol, S. Synthesis of plant oil-based amide copolymethacrylates and their use as viscosity index improvers. *Eur. Polym. J.* **2018**, *109*, 435–446. [CrossRef]
24. Rathika, S.; Raghavan, P.S. Influence of polyvinyl palmitate copolymer as viscosity index improvers For lube. *Mater. Today Proc.* **2019**, *16*, 978–986. [CrossRef]
25. Esfea, M.H.; Aranib, A.A.A.; Esfandeh, S. Improving engine oil lubrication in light-duty vehicles by using of dispersing MWCNT and ZnO nanoparticles in 5W50 as viscosity index improvers (VII). *Appl. Therm. Eng.* **2018**, *143*, 493–506. [CrossRef]
26. Méheust, H.; Le Meins, J.-F.; Brûlet, A.; Sandre, O.; Grau, E.; Cramail, H. Fatty-acid based comb copolyesters as viscosity Index improvers in lubricants. *Eur. Polym. J.* **2022**, *181*, 111674. [CrossRef]
27. Sparnacci, K.; Frison, T.; Podda, E.; Antonioli, D.; Laus, M.; Notari, M.; Assanelli, G.; Atzeni, M.; Merlini, G.; Pó, R. Core-crosslinked star copolymers as viscosity index improvers for lubricants. *ACS Appl. Polym. Mater.* **2022**, *4*, 8722–8730. [CrossRef]
28. Khalafvandi, S.A.; Pazokian, M.A.; Fathollahi, E. The investigation of viscometric properties of the most reputable types of viscosity index improvers in different lubricant base oils: API groups I, II, and III. *Lubricants* **2022**, *10*, 6. [CrossRef]
29. Tseregounis, S.I.; McMillan, M.L.; Olree, R.M. Engine oil effects on fuel economy in GM vehicles-separation of viscosity and friction modifier effects. In *SAE Technical Paper Series-Fuel Economy and Wear Performance on Engine Oils*; Society of Automotive Engineers, Inc.: Warrendale, PA, USA, 1998; p. 982502.
30. Zhang, D.; Li, Z.; Wei, X.; Wang, L.; Xu, J.; Liu, Y. Study tribological properties of MoDTC and its interactions with metal detergents. *J. Tribol.* **2020**, *142*, 122201. [CrossRef]
31. Pirouz, S.; Duhamel, J. Using pyrene excimer fluorescence to probe the interactions between viscosity index improvers and waxes present in automotive oil. *Macromolecules* **2017**, *50*, 2467–2476. [CrossRef]
32. Gholami, K.; Frasca, F.; Duhamel, J. Probing the interactions between pour point depressants (PPDs), viscosity index improvers (VIIs), and wax in octane using fluorescently labeled PPDs *Can. J. Chem.* **2022**, *100*, 688–696. [CrossRef]
33. Tan, L.; Huang, T.; Ma, J.; Li, J.; Song, Z.; Zhou, X.; Tang, Y.; Yang, L.; Zeng, X. Study on the effect of end group on the anti-corrosion behaviour of polyether derivatives. *J. Mol. Liq.* **2022**, *347*, 117991. [CrossRef]
34. Guo, Y.; Li, J.; Zhou, X.; Tang, Y.; Zeng, X. Formulation of lyotropic liquid crystal emulsion based on natural sucrose ester and its tribological behaviour as novel lubricant. *Friction* **2022**, *10*, 1879–1892. [CrossRef]
35. *GB/T 265*; Petroleum Products-Determination of Kinematic Viscosity and Calculation of Dynamic Viscosity. General Administration of Quality Supervision, Inspection and Quarantine of the People's Republic of China: Beijing, China, 1998.
36. *GB/T 1995*; Petroleum Products-Calculation of Viscosity Index. The State Bureau of Quality and Technical Supervision: Beijing, China, 1998.

37. *GB/T 3535;* Petroleum Products-Determination of Pour Point. General Administration of Quality Supervision, Inspection and Quarantine of the People's Republic of China: Beijing, China, 2006.
38. *GB/T 3536;* Petroleum Products-Determination of Flash and Fire Points-Cleveland Open Cup Method. General Administration of Quality Supervision, Inspection and Quarantine of the People's Republic of China: Beijing, China, 2008.
39. *GB/T 6538;* Determination of Apparent Viscosity of Engine Oils-Using the Cold-Cranking Simulator. The State Bureau of Quality and Technical Supervision: Beijing, China, 2022.
40. *SH/T 0562;* Standard Test Method for Determination of Yield Stress and Apparent Viscosity of Engine Oils at Low Temperature. National Energy Administration: Beijing, China, 2013.
41. *SH/T 0703;* Standard Test Method for Measuring Apparent Viscosity at High-Temperature and High-Shear Rate by Multicell Capillary Viscometer. National Energy Administration: Beijing, China, 2021.
42. *SH/T 0059;* Standard Test Method for Evaporation Loss of Lubricating Oils by the Noack Method. National Energy Administration: Beijing, China, 2011.
43. *GB/T 12579;* Determination of Foaming Characteristics of Lubricating Oils. General Administration of Quality Supervision, Inspection and Quarantine of the People's Republic of China: Beijing, China, 2002.

lubricants

Article

Start-Up Process of High-Speed Micro-Grooved Pumping Seal for New Energy Vehicles

Hanqing Chen [1,2,*], Ruqi Yan [3], Xianzhi Hong [4], Xin Bao [4] and Xuexing Ding [1,*]

1 School of Petrochemical Engineering, Lanzhou University of Technology, Lanzhou 730050, China
2 School of Mechanical Engineering, Lanzhou Petrochemical University of Vocational Technology, Lanzhou 730060, China
3 School of Chemistry and Chemical Engineering, Lanzhou Jiaotong University, Lanzhou 730070, China; yanruqima@126.com
4 Chengdu Yitong Seal Co., Ltd., Chengdu 610100, China; hongxianzhi@cdytseal.com (X.H.); baoxin@cdytseal.com (X.B.)
* Correspondence: 181080706002@lut.edu.cn (H.C.); dingxxseal@126.com (X.D.); Tel.: +86-188-9348-6963 (X.D.)

Abstract: With the growing global demand for clean energy, new energy vehicles are a key focus in the automotive industry. This paper investigates the micro-grooved pumping seal used in such vehicles, using a custom Python computational programme to study the start-up behaviour of a non-contact oil–gas two-phase micro-grooved seal. The research explores the balance of forces during start-up, employing fractal theory for surface contact force calculations and solving the two-phase laminar Reynolds equation by the finite difference method. The results show that high-speed micro-grooved seals perform well under typical conditions for new energy vehicles. When film thickness is below a critical value, fractal dimension and characteristic length influence the initial thickness. Above the critical value, film thickness increases non-linearly with rotational speed, whereas the leakage rate decreases linearly. Critical rotational speed decreases non-linearly with the oil–gas ratio, peaking at an oil–gas ratio of 0.06. Both critical speed and leakage rate increase linearly and non-linearly with pressure and temperature, respectively. The study highlights the boundary-line where leakage transitions to pumping, providing valuable guidance for optimising seal design in new energy vehicles.

Keywords: micro-grooved pumping seal; micro-convex body contact; oil–gas two-phase flow; start-up process; sealing performance; new energy vehicle

1. Introduction

With the continuous growth in global demand for clean energy, new energy vehicles, including pure electric and hybrid vehicles, have become a primary focus of development in the automotive industry. As a core component of these vehicles, the electric drive system significantly impacts overall performance, efficiency, and reliability. However, the high-speed operation and complex working conditions of the electric drive system impose stringent requirements on sealing technology. In extreme working environments, such as high rotational speeds, two-phase fluids, and bidirectional rotation, traditional sealing technologies show significant shortcomings in preventing lubricant leakage, reducing friction losses, and maintaining system stability.

In recent years, micro-grooved pumping seal technology has gradually emerged to address these challenges. This technology, based on the fluid dynamic pressure effect, is a non-contact sealing method suitable for high-speed rotating equipment. Its unique micro-groove design generates a dynamic pressure effect at the sealing interface, significantly enhancing the seal's load-bearing capacity and overall reliability. Compared to traditional rubber skeleton oil seals [1,2] and polytetrafluoroethylene (PTFE) oil seals [3–5], micro-grooved pumping seals perform exceptionally well under conditions of high rotational speed, heavy loads, and bidirectional rotation, showing great potential in the electric drive systems of new energy vehicles. Additionally, since electric motors typically operate at extremely high rotational speeds (progressing towards exceeding 40,000 rpm [6]), the stability and anti-reversal capability of the seal during start-up are crucial factors that determine the motor's lifespan and performance. Traditional sealing technologies have weaker anti-reversal capabilities, whereas micro-grooved pumping seals, with their bidirectional micro-groove design on the sealing surface, provide a more stable fluid dynamic pressure effect during start-up, significantly enhancing anti-reversal capability. The research and application of this technology not only help improve the reliability and lifespan of seals but also effectively reduce lubricant leakage and environmental pollution, representing the future direction of sealing technology for new energy vehicles. However, the dynamic behaviour of the sealing system during the start-up process directly affects its reliability and long-term stable operation. Therefore, in-depth research on the start-up process of micro-grooved pumping seals, particularly the changes in sealing performance during this phase, has become a critical problem that needs to be addressed.

When examining the flow field characteristics during the start-up process of mechanical seals, the morphology and contact behaviour of the sealing surface have a direct impact on fluid flow patterns. Greenwood and Williamson [7] proposed a classic model for studying the contact of nominally flat surfaces, providing a foundational theory for analysing mechanical seals and lubrication interfaces. Mandelbrot [8] suggested that fractal geometry could effectively describe complex random phenomena in nature, such as terrain undulations and coastline shapes, offering a new perspective on contact mechanics. Johnson [9] established a basic framework of contact theory, providing classical theoretical support for contact problems in lubrication and tribology. Majumdar and Bhushan [10] explored the role of fractal geometry in roughness characterisation and surface contact mechanics, proposing new methods for analysing surface roughness. Based on fractal geometry, Majumdar and Bhushan [11] developed a fractal model of elastic-plastic contact between rough surfaces, demonstrating its effectiveness in describing the relationship between surface roughness and contact behaviour. Wang and Komvopoulos [12] studied the interfacial temperature distribution under slow sliding conditions using fractal theory, concluding that fractal models can effectively describe thermal transfer behaviour at contact interfaces, and revealing the impact of thermal coupling on friction performance. Dong and Zhang [13] refined the M-B elastic-plastic contact model, finding that it more accurately reflects actual operating conditions when describing the elastoplastic contact behaviour of rough surfaces. Ge and Zhu [14] highlighted the significant potential of fractal theory in tribology for optimising friction behaviour under complex conditions. Bhushan [15,16] further investigated the application of fractal geometry in surface roughness characterisation, evaluated the measurement capabilities of different experimental techniques for real contact areas, and proposed new experimental methods to improve measurement accuracy. Wei et al. [17] developed a sliding friction surface contact mechanics model based on fractal theory, analysing the impact of rough surfaces on friction performance and suggesting the optimisation of friction interface design through fractal models. Ding et al. [18] constructed a rough surface contact model based on fractal theory, analysing the impact of different base lengths on sealing interface performance and proposing design solutions to enhance sealing performance. Zhao et al. [19] proposed a prediction model for mechanical seal leakage rate and film thickness based on fractal contact theory, suggesting that optimising these parameters can significantly improve the reliability and lifespan of seals.

In addition to surface contact characteristics, lubrication and dynamic pressure effects at the microscale are crucial factors influencing sealing performance. Fukui and Kaneko [20], through their analysis of the lubrication mechanism of ultra-thin gas films, proposed a generalised lubrication equation, which they believe is significant for the design of microscale lubrication systems. Gu [21] highlighted that the lift-off phenomenon and lift characteristics of the sealing face in mechanical face seals significantly affect the stability and lifespan of the seal. Li et al. [22] emphasised that the opening characteristics of dry gas seals are critical for enhancing sealing effectiveness, particularly during the start-up process, where the lift-off phenomenon of the sealing face directly impacts the control of working pressure. Fan et al. [23] investigated the start-up process of dry gas seals under steam lubrication conditions and found that steam lubrication can effectively improve the start-up performance of seals, particularly in high-humidity environments. Sun et al. [24] analysed the impact of seal ring material properties and surface topography on the dynamic contact characteristics of dry gas seals during start/stop phases, providing recommendations for optimising sealing performance.

While fractal geometry models and contact mechanics have made significant progress in describing microscopic surface contact behaviour, existing models are often applied under steady-state conditions, and their impact on the dynamic response of micro-grooved pumping seals during the start-up phase has not been fully explored. As a result, there is still a lack of research combining fluid dynamic pressure effects with fractal contact models during the start-up process, particularly in the transient behaviour analysis of seals under high rotational speeds, two-phase flow, and bidirectional rotation conditions. This gap represents a critical bottleneck limiting the broader application of this technology.

Considering this background, this study focused on analysing the changes in the sealing performance of high-speed micro-grooved pumping seals for new energy vehicles during the start-up process. Before the separation of the end faces, the research applied tribology and contact mechanics as foundational theories, with fractal geometry models used to describe the contact behaviour of microscopic surfaces. This approach explored the influence of surface contact microstructures and contact characteristics on sealing performance. After the separation of the end faces, the study employed the compressible steady-state two-phase laminar Reynolds equation, incorporating real gas effects. The virial and Lucas equations were utilised to describe changes in real gas effects and viscosity, respectively, while the oil–gas ratio was introduced to determine the equivalent density and equivalent viscosity. Utilising numerical simulations, the study investigated the dynamic pressure effects under transient start-up conditions and their impact on sealing performance, addressing gaps in existing research. These innovative investigations will not only enhance the understanding of the start-up process of micro-grooved pumping seals but also provide essential theoretical foundations and practical insights for the sealing design of electric drive systems in future new energy vehicles.

2. Theoretical Framework

2.1. Geometric Framework

The micro-grooved pumping seal structure for new energy vehicles is shown in Figure 1. The rotating ring is made of silicon carbide, while the stationary ring is crafted from graphite, with grooves etched onto the rotating ring. The rotating ring is secured within the rotating ring seat, which is positioned relative to the shaft using a pin and sealed with an O-ring. On the opposite side, the stationary ring is connected to the sealing cover via a bellows (wave spring), which serves both as an elastic element and a seal. The sealing cover is further sealed to the sealing chamber with a gasket. During operation, the micro-grooved pumping seal allows the oil–gas mixture to enter from the low-pressure side, where it is pumped through the dynamic pressure grooves on the sealing face to the high-pressure side. This creates an ultra-thin oil–gas two-phase film on the sealing face, effectively achieving zero oil and gas leakage. When the system is shut down, the sealing faces close, preventing lubricant leakage and ensuring a tight seal. This is a new type of seal [25].

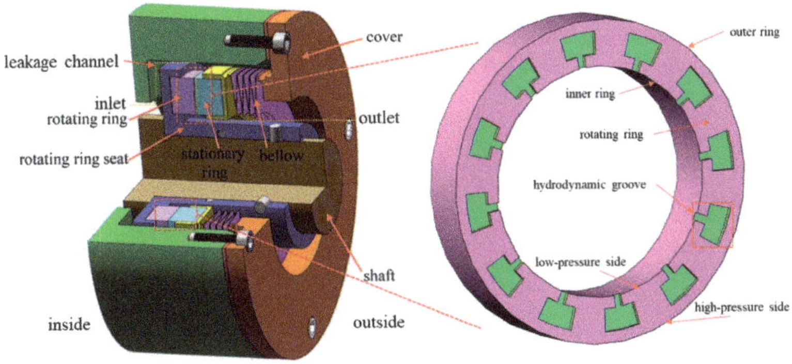

Figure 1. Schematic of proposed micro-grooved pumping seal for new energy vehicles.

2.2. Force Balance During the Start-Up Process

As shown in Figure 2, various forces, such as spring force, fluid static pressure, fluid dynamic pressure, and surface contact force, act on the sealing rings during the start-up of this new type of seal for new energy vehicles. These forces balance each other, collectively determining the relative position of the two sealing rings. Regardless of whether the sealing faces are rotating relative to each other or stationary, the closing force is always present. This force is generated partly by the pressure acting on the non-contacting back area of the sealing ring and partly by the spring pressure. At the initial stage of start-up, or when the two sealing faces are stationary, the end faces of the rotating and stationary rings, which have a certain surface roughness, experience micro-convex body contact. The interaction between these micro-convex bodies generates the surface contact force. During the start-up process, in addition to the surface contact force, fluid dynamic pressure is generated by the oil–gas mixture under the influence of the hydrodynamic grooves. At this stage, the surface contact force and fluid dynamic pressure together create the opening force. As the rotational speed increases, the two sealing faces gradually separate due to the fluid dynamic pressure, and eventually, the opening force is entirely sustained by the fluid dynamic pressure, referred to as the film opening force.

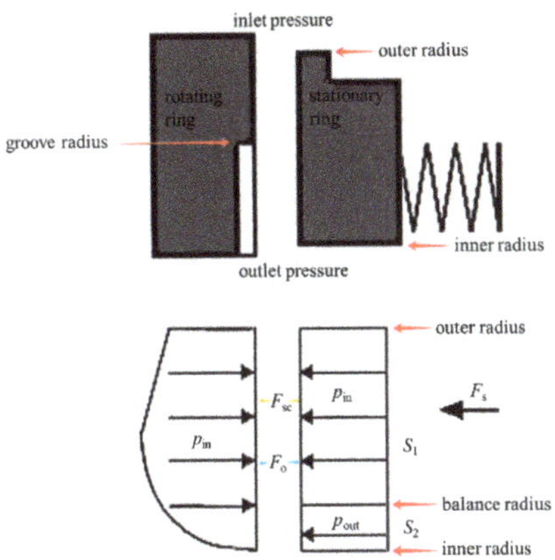

Figure 2. Schematic of the balance of forces.

The closing force is composed of the fluid static pressure and the spring force acting on the sealing ring. Typically, the spring force varies with the compression length in its calculation formula. However, since the compression length in this study is at the micron level, the variation in spring force is negligible and can be considered constant. As a result, the spring force is treated as a fixed value in this study.

The formula for calculating the closing force is:

$$F_c = S_1 \cdot (p_{in}) + S_2 \cdot (p_{out}) + F_s \tag{1}$$

where S_1 is the high-pressure side surface area of the back of the sealing ring (in m^2); p_{in} is the inlet pressure (i.e., the gas pressure at the outer radius of the sealing ring, the high-pressure side, in Pa); S_2 is the low-pressure side surface area of the back of the sealing ring (in m^2); p_{out} is the outlet pressure (i.e., the gas pressure at the inner radius of the sealing ring, the low-pressure side, in Pa); and F_s is the spring force (N).

During the start-up process, the sum of the surface contact force and the film opening force equals the closing force, but they have an inverse relationship. As the film opening force increases, the surface contact force decreases, and vice versa. The force balance relationship during the start-up process is as follows:

$$F_c = F_{sc} + F_o \tag{2}$$

where F_c is the closing force (N), F_{sc} is the surface contact force (N), and F_o is the film opening force (N).

2.3. Critical States During the Start-Up Process

During the start-up process, initially, the two sealing faces are at rest relative to each other. As the rotational speed gradually increases, the faces transition from initial contact and friction to being finally separated by the fluid's hydrodynamic pressure. The "critical state" refers to the boundary condition during start-up, where the faces transition from contact to non-contact.

2.3.1. Critical Film Thickness h_c

The critical film thickness is a dimensional threshold. When the film thickness is below this critical film thickness, it falls under mixed friction [21]; when it exceeds the critical film thickness, it transitions to fluid friction, which, in this study, is also referred to as oil–gas two-phase flow lubrication. When the film thickness is less than the critical film thickness, the rotating and stationary end faces have not fully separated. At this stage, both oil–gas two-phase lubrication and surface contact are present. Therefore, when the film thickness is below the critical film thickness, the film opening force consists of the fluid dynamic pressure from the oil–gas two-phase film and the surface contact force from the micro-convex bodies on the rough surfaces of the sealing faces. Gu [21] referred to the critical film thickness as the contact film thickness, which represents the critical point wherein mixed friction transitions to fluid friction. It can be expressed as:

$$h_c = 3.75 \cdot \sqrt{{R_{a1}}^2 + {R_{a2}}^2} \tag{3}$$

where R_{a1} and R_{a2} are the surface roughness of the harder and softer materials, respectively (in m).

According to the "specification for dry gas seal" [26], $R_{a1} \leq 0.1$ μm and $R_{a2} \leq 0.2$ μm; thus, the maximum critical film thickness is $h_c \approx 0.85$ μm under these conditions.

2.3.2. Critical Rotational Speed n_c

The critical rotational speed represents the threshold for force balance: below this speed, the surface contact force and film opening force together balance the closing force, while above it, only the film opening force balances the closing force. When the gap between the rotating and stationary rings reaches the critical film thickness, the rotational speed of the rotating ring causes the oil–gas two-phase film to generate a film opening force that balances the closing force, resulting in a surface contact force of zero. This speed is known as the critical rotational speed. Once the rotational speed exceeds the critical rotational speed, the film opening force increases. When the film opening force surpasses the closing force, the film thickness also increases, exceeding the critical film thickness. As the film thickness increases, the film opening force decreases. When the film opening force decreases sufficiently to balance the closing force, the film thickness stabilises or fluctuates slightly around the new equilibrium value. This is a dynamic equilibrium process, simplified here as a quasi-static equilibrium process. Thus, as the rotational speed continues to increase, the rotating and stationary rings reach new equilibrium positions, with the film thickness continually increasing to maintain the balance between the film opening force and the closing force. With a constant closing force, the critical rotational speed is influenced by factors such as inlet pressure, oil–gas ratio, and temperature. The purpose of this study is to explore how these factors affect the critical rotational speed and to optimise performance parameters during the start-up process of this new type of seal for new energy vehicles.

2.4. *Rough Surface Contact*

2.4.1. Surface Contact Process

Due to the roughness of the sealing end faces, when the device is stationary and the two sealing faces are in contact, or during the initial stage of start-up, a portion of the surfaces remains in contact. The contact of rough surfaces can be described using fractal contact theory. Since the rotating and stationary rings of this new type of seal involve a combination of a hard and a soft ring, the contact analysis of the sealing friction interface can be simplified to the interaction between an ideal rigid smooth surface and a rough surface, as illustrated in Figure 3. Under the applied load F_{sc}, the rigid surface moves in the negative direction along the z-axis, and the micro-convex bodies sequentially undergo plastic, elastoplastic, and elastic contact states [27].

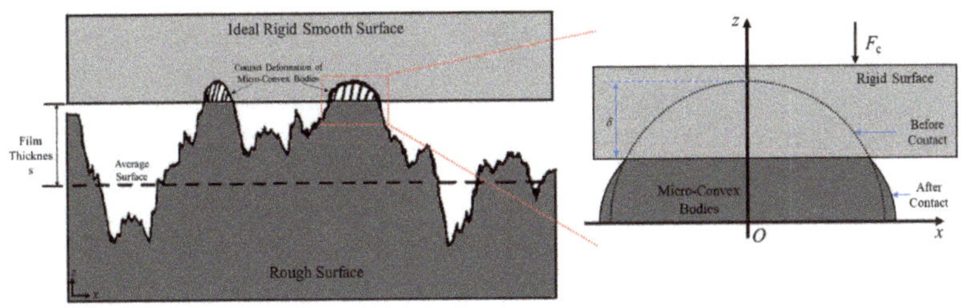

Figure 3. Contact deformation diagram of rough surface micro-convex body.

The normal deformation of the micro-convex bodies can be represented by the normal displacement, which is related to the maximum contact area of the micro-convex bodies on the contact surface. This relationship can be expressed as follows:

$$\begin{cases} \delta = G^{(D-1)} \cdot a_1{}^{(2-D)/2} \\ a_1 = \frac{2-D}{D} \cdot \psi^{(D-2)/2} \cdot A_r \\ \psi = 5.4532 \exp\left(\frac{-D}{0.62782}\right) + 1.499 \end{cases} \tag{4}$$

where δ is the normal displacement (in m); G is the characteristic length scale (in m); D is the fractal dimension (dimensionless); a_1 is the maximum contact area of the micro-convex bodies (in m^2); ψ is the area expansion coefficient (dimensionless); and A_r is the real contact area (in m^2).

The analysis clearly demonstrates that comprehending the correlation between the real area of contact and the surface contact force enables the determination of the relationship between the surface contact force and the normal displacement.

2.4.2. Surface Contact Force

The interaction force between rough surfaces is referred to as the surface contact force. It can be calculated using rough surface contact models based on statistical or fractal parameters. According to the fractal contact theory, the relationship between the real contact area and the surface contact force is as follows [14]:

$$F_{sc} = \begin{cases} \frac{4}{3}\sqrt{\pi}E \cdot A_a \cdot {G^*}^{D-1} \cdot g_1(D) \cdot \psi^{\frac{(D-2)^2}{4}} \cdot {A_r^*}^{\frac{D}{2}} \cdot \left[\left(\frac{2-D}{D} \cdot \psi^{\frac{D-2}{2}} \cdot A_r^* \right)^{\frac{3-2D}{2}} - {a_c^*}^{\frac{3-2D}{2}} \right] & \\ \qquad + K \cdot \sigma_y \cdot A_a \cdot g_2(D) \cdot \psi^{\frac{(D-2)^2}{4}} \cdot {A_r^*}^{\frac{D}{2}} \cdot {a_c^*}^{\frac{2-D}{2}} & D \neq 1.5 \\ 3^{-\frac{3}{4}}\sqrt{\pi}E \cdot A_a \cdot {G^*}^{\frac{1}{2}} \cdot \psi^{\frac{1}{16}} \cdot {A_r^*}^{\frac{3}{4}} \cdot \ln\left(\frac{A_r^*}{3\psi^{\frac{1}{4}} \cdot a_c^*} \right) + 3^{\frac{1}{4}} K \cdot \sigma_y \cdot A_a \cdot \psi^{\frac{1}{16}} \cdot {A_r^*}^{\frac{3}{4}} \cdot {a_c^*}^{\frac{1}{4}} & D = 1.5 \end{cases} \tag{5}$$

where:

$$\begin{cases} g_1(D) = \frac{D}{3-2D} \cdot \left(\frac{2-D}{D} \right)^{\frac{D}{2}} \\ g_2(D) = \left(\frac{D}{2-D} \right)^{\frac{2-D}{2}} \\ a_c = G^2 \cdot \left(\frac{\pi E^2}{225\sigma_y^2} \right)^{\frac{1}{D-1}} \\ E = \left[\frac{1-v_1^2}{E_1} + \frac{1-v_2^2}{E_2} \right]^{-1} \\ G^* = \frac{G}{\sqrt{A_a}} \\ {A_r}^* = \frac{A_r}{A_a} \\ a_c^* = \frac{a_c}{A_a} \end{cases} \tag{6}$$

where σ_y is the yield strength of the softer material (in Pa); ν_1 and ν_2 are the Poisson's ratios of the harder and softer materials, respectively (dimensionless); E_1 and E_2 are the elastic moduli of the harder and softer materials, respectively (in Pa); E is the equivalent elastic modulus (in Pa); A_a is the nominal contact area (in m^2); a_c is the critical contact area (in m^2); and $g_1(D)$, $g_2(D)$, G^*, A_r^*, a_c^*, and K are detailed in reference [14].

2.5. Performance Parameters of Oil–Gas Two-Phase Film

2.5.1. Equation for Pressure Control

The fluid within the seal was presumed to be a Newtonian fluid exhibiting a zero-pressure gradient in the direction of film thickness. The deformation of the seal end surface was considered minimal, and temperature changes inside the flow field were not taken into account. Therefore, the steady-state Reynolds equation for the laminar flow field in this new type of seal can be given as follows [28,29]:

$$\frac{1}{r} \cdot \frac{\partial}{\partial \theta}\left(\frac{\rho_e \cdot h^3}{12 \cdot \eta_e} \cdot \frac{\partial p}{\partial \theta} \right) + \frac{\partial}{\partial r}\left(\frac{r \cdot \rho_e \cdot h^3}{12 \cdot \eta_e} \cdot \frac{\partial p}{\partial r} \right) = r \cdot \omega \cdot \frac{\partial}{\partial \theta}(\rho_e \cdot h) \tag{7}$$

where p, r, θ, ω, h, ρ_e, η_e, and the detailed calculation process of Equation (7) are provided in reference [25].

2.5.2. Steady-State Performance Metrics

By resolving the Equation (7), the opening force (F_o) and leakage rate (Q_m) can be obtained as follows:

$$F_o = \int_0^{2\pi} \int_{r_i}^{r_o} p \cdot r dr d\theta \tag{8}$$

and the leakage mass rate is:

$$Q_m = \int_0^{2\pi} \int_0^{h} \rho_e \cdot u_r \cdot r dz d\theta \tag{9}$$

3. Results and Discussion

The pressure control problem was numerically solved using the finite difference method, taking into account the impacts of real gas as well as the equivalent density and viscosity of the oil–gas two-phase mixture. The pumping mechanism and sealing properties were further analysed and commented. The geometric configuration and specifications of this new type of seal (which is an inverted trapezoidal groove in this work), as well as the operational circumstances and material characteristics, are presented in Figure 4, Table 1, Table 2, and Table 3, respectively. Figure 5 is the technical roadmap, where the "fluid dynamic pressure model" is based on reference [25].

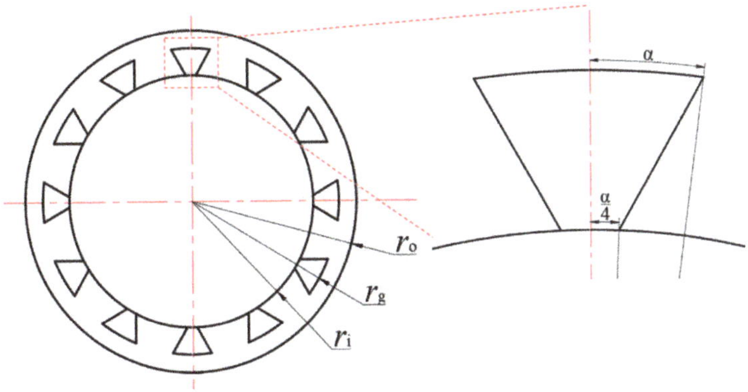

Figure 4. Schematic of dynamic ring size parameters.

Table 1. Geometric parameters of high-speed micro-groove pumping seal for new energy vehicles.

Parameter	Value
Outer radius r_o/mm	38
Inner radius r_i/mm	28
Number of grooves N_g	12
Groove depth h_g/μm	5
Groove root radius r_g/mm	34
Expanding angle of groove α/°	$180/(2 \cdot N_g)$

Table 2. Operating conditions parameters of high-speed micro-groove pumping seal for new energy vehicles.

Parameter	Value
Rotational velocity n/rpm	0~20,000
Inlet gauge pressure p_i/kPa	0~300
Outlet gauge pressure p_o/kPa	0
Inlet temperature T_i/°C	40~130
Outlet temperature (environmental) T_o/°C	20
Oil–gas ratio c	0.01~0.1
Spring force F_s/N	49

The parameters used in the calculation are all listed in Tables 1 and 2, which include p_i = 0.1 MPa, T_i = 100 °C, n = 20000 rpm, and an oil–gas ratio of 0.03, employing a Python programme. A Python programme I developed was employed, which contains all the detailed information regarding grid size, boundary conditions, and convergence criteria [25]. Figure 6 demonstrates the effect of grid density, utilising a grid count of 19,716 for this analysis.

Table 3. Material characteristic parameters of high-speed micro-groove pumping seal for new energy vehicles.

Parameter	Silicon Carbide Ring	Graphite Ring
Elastic modulus, E/GPa	400	20
Poisson's ratio, ν	0.14	0.24
Yield strength, σ_y/MPa	—	200
Fractal dimension D	—	1.6
Characteristic length scale G/m	—	6.7×10^{-9}

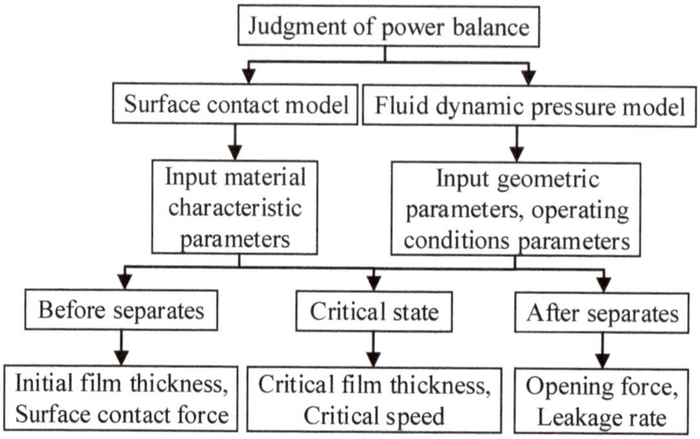

Figure 5. Technical roadmap.

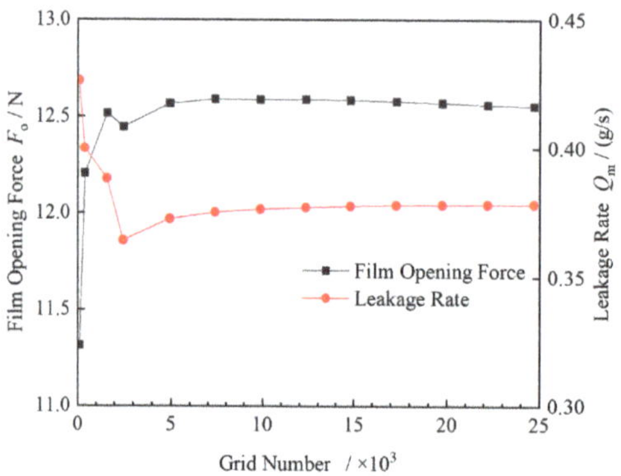

Figure 6. The effect of grid density.

3.1. Relationship Between Surface Contact Force and Film Opening Force During the Start-Up Process

By substituting the parameters from Tables 1 and 3 into Equations (4) and (5), the variations in surface contact force, film opening force, and film thickness were calculated. The blue dashed auxiliary lines in Figure 7 indicate the initial film thickness and critical film thickness under the operating conditions of this study.

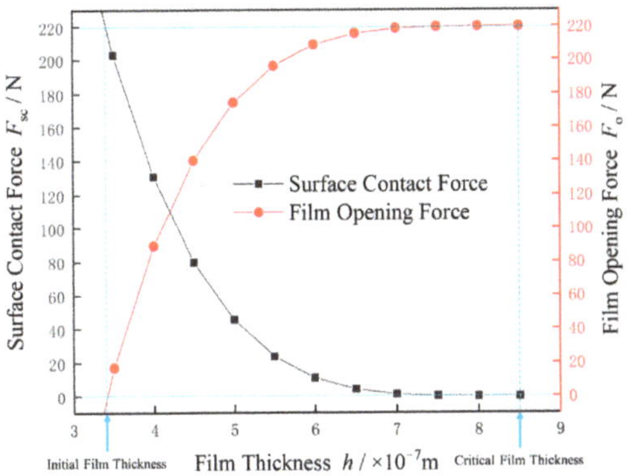

Figure 7. Relationship between surface contact force, film opening force, and film thickness.

According to Figure 7, the surface contact force decreases non-linearly as the film thickness increases. Conversely, as the film thickness decreases from the critical film thickness to zero, the surface contact force increases, and the rate of this increase also rises. According to Equation (2), the sum of the surface contact force and the film opening force always equals the closing force. When the surface contact force equals the closing force, the film opening force is zero, and the two sealing end faces are relatively stationary. At this point, the film thickness corresponds to the initial film thickness. When the surface contact force is zero and the film opening force equals the closing force, the rotational speed corresponds to the critical rotational speed, and the corresponding film thickness is the critical film thickness.

3.2. *Influence of Fractal Parameters on Initial Film Thickness*

The distance between the critical film thickness and the initial film thickness represents the maximum normal displacement of the micro-convex bodies. By varying the fractal dimension and characteristic length scale in the fractal contact model, it is conceivable to comprehend how variations in surface microtopography affect the initial film thickness.

3.2.1. Influence of Fractal Dimension

Using the parameters from Tables 1–3, and keeping the characteristic length scale constant, the relationship between the fractal dimension and the initial film thickness was calculated. As shown in Figure 8, the blue dashed auxiliary lines indicate the initial film thickness for various fractal dimensions.

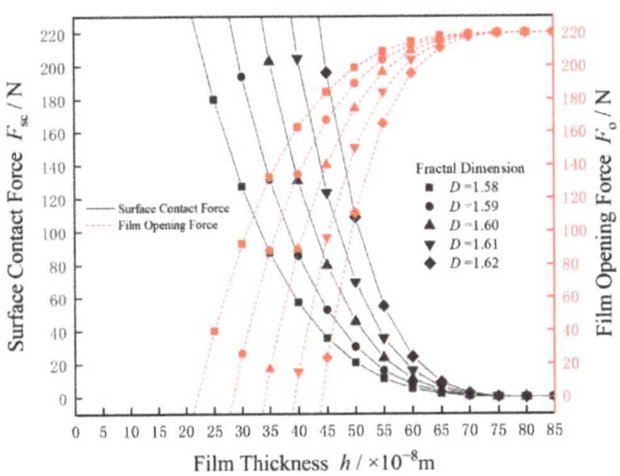

Figure 8. Initial film thickness under different fractal dimensions.

Figure 8 shows that the initial film thickness increases with the fractal dimension, but the rate of increase gradually diminishes. Alternatively, this can be interpreted as a decrease in the maximum normal displacement of the micro-convex bodies. This occurs because, as the fractal dimension increases, the self-similarity of the microstructure enhances significantly, making the surface structure more complex and detailed. The number of micro-convex bodies increases, expanding the effective contact area and dispersing the normal force. Consequently, the increased complexity of the micro-convex body shapes reduces the concentration of normal displacement, thereby improving the average deformation resistance of the micro-convex bodies. This leads to superior mechanical performance of the overall structure.

3.2.2. Influence of Characteristic Length Scale

Using the parameters from Tables 1–3, and keeping the fractal dimension constant, the relationship between the characteristic length scale and the initial film thickness was calculated. As shown in Figure 9, the blue dashed auxiliary lines indicate the initial film thickness for various characteristic length scales.

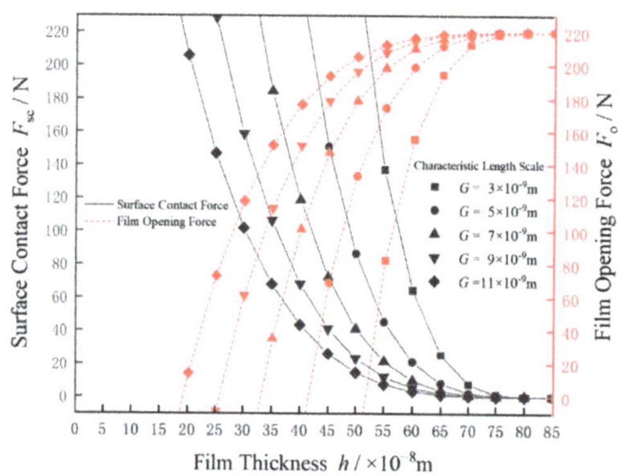

Figure 9. Initial film thickness under different characteristic length scales.

Figure 9 shows that the initial film thickness decreases as the characteristic length scale increases, with the rate of decrease gradually diminishing. This indicates that the maximum normal displacement of the micro-convex bodies increases. As the characteristic length scale increases, larger protrusions and depressions on the surface become more pronounced, leading to increased surface roughness. Additionally, the leakage channels and porosity between the contact surfaces may expand, affecting sealing performance and increasing the risk of leakage. Although the closing force remains constant, changes in surface topography cause uneven pressure distribution within the contact area, resulting in more significant local deformation of some micro-convex bodies during contact. Consequently, the normal displacement of the micro-convex bodies becomes larger, leading to a smaller initial film thickness.

The observations from the previous two sections indicate that changes in the fractal dimension and characteristic length scale significantly impact the initial film thickness during the start-up process of this new type of seals for new energy vehicles. By optimising the micro-structure and reducing contact porosity, the sealing performance of this new type of seal can be effectively enhanced, both when stationary and before separation. This research provides important guidance for material selection and surface treatment processes, thereby reducing the risk of leakage during the start-up process.

3.3. Analysis of Rotational Speed, End Face Gap, and Leakage After Separation

Utilizing the data from both Tables 1 and 2 for an input with a gauge pressure of 100 kPa, a temperature setting of 100 °C, and an oil–gas ratio of 0.03, the fluctuations in film thickness and leakage rate were computed as the rotating speed progressively increased from 0 to 20,000 r/min. The findings are illustrated in Figure 10.

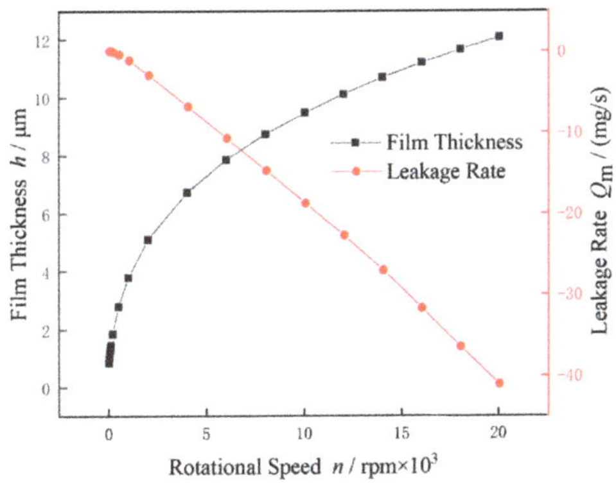

Figure 10. Film thickness and leakage rate at different rotational speeds.

Figure 10 shows that the film thickness increases non-linearly with rotational speed, rising rapidly at first and then more gradually. This trend indicates that the stiffness of the oil–gas two-phase film continuously decreases. Although increasing the rotational speed enhances the film opening force, the force balance relationship dictates that the film opening force is always balanced by the closing force, both of which remain constant. Therefore, as the rotational speed increases, the film thickness continues to increase until the forces reach a new balance. The continuously increasing film thickness, combined with the constant film opening force, results in a decrease in stiffness.

The leakage rate changes almost linearly with the rotational speed. It is important to note that the seal chamber serves as the inlet, while the environment acts as the outlet. When the sign of the leakage rate is positive, it indicates that the leakage direction flows from the inlet to the outlet, meaning that oil and gas are leaking from the sealed leakage channel to the outside, i.e., oil–gas leaks from the seal chamber into the environment. Conversely, a negative leakage rate means the flow is from the outlet to the inlet, i.e., from the environment into the seal chamber, referred to as the pumping rate. Therefore, as the rotational speed increases, the sealing system transitions from leakage to pumping. The rotational speed at which this transition occurs is called the boundary-line speed between leakage and pumping, which is approximately 102 r/min under these operating conditions.

3.4. Influence of Operating Condition Changes on Critical States

By altering various operating parameters, changes in the critical rotational speed were observed when the critical film thickness was set to 0.85 μm. Additionally, the performance parameters at this critical rotational speed were examined, with particular emphasis on the leakage rate.

3.4.1. Influence of Oil–Gas Ratio on Start-Up Performance

Applying the data from both Tables 1 and 2 for an inlet with a gauge pressure of 100 kPa, a temperature setting of 100 °C, and an oil–gas ratio incrementally increasing from 0.01 to 0.1, the variations in critical rotational speed and leakage rate were calculated. The results are presented in Figure 11.

Figure 11 shows that the critical rotational speed decreases non-linearly as the oil–gas ratio increases. This is because, as the oil–gas ratio increases, the equivalent density and viscosity of the oil–gas two-phase mixture also increase. As a result, a lower rotational speed is sufficient to generate the same film opening force and achieve the same critical film thickness.

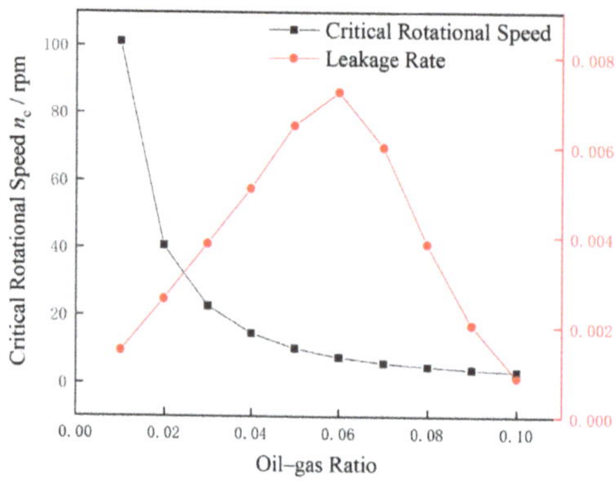

Figure 11. Critical rotational speed and leakage rate at different oil–gas ratios.

At different critical states corresponding to various oil–gas ratios, the leakage rate initially increases and then decreases non-linearly, peaking at an oil–gas ratio of approximately 0.06. According to previous research [25], when the rotational speed is constant, an increase in the oil–gas ratio increases the pumping rate, thereby reducing the leakage rate. Similarly, when the oil–gas ratio is constant, an increase in rotational speed decreases the leakage rate. Figure 11 highlights a special operating point during the start-up process: when the oil–gas ratio is approximately 0.06, the leakage rate is relatively high, either during start-up or during sustained low-speed operation after start-up. This suggests that in seal design, it is advisable to avoid an oil–gas ratio near 0.06. If an oil–gas ratio near 0.06 offers significant advantages in other performance aspects, it is crucial to minimise the number of start-stop cycles, accelerate quickly after start-up to reduce low-speed operation time, and limit the minimum rotational speed after start-up to effectively reduce the leakage rate.

3.4.2. Influence of Inlet Gauge Pressure on Start-Up Performance

Using the parameters from Tables 1 and 2, with an oil–gas ratio of 0.03, a temperature of 100 °C, and an inlet gauge pressure incrementally increasing from 0 to 300 kPa, the variations in critical rotational speed and leakage rate were calculated. The results are presented in Figure 12.

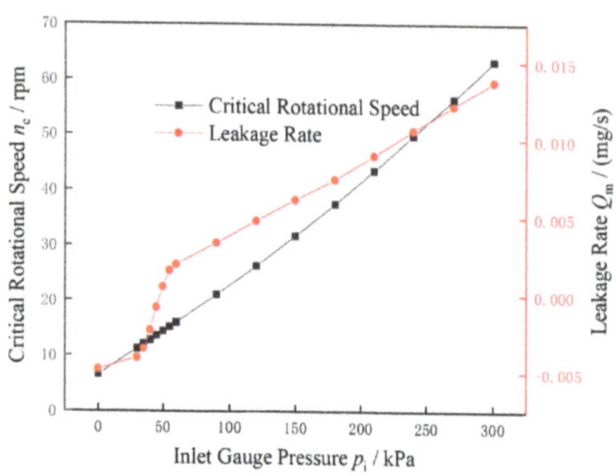

Figure 12. Critical rotational speed and leakage rate at different inlet gauge pressures.

Figure 12 shows that the critical rotational speed increases approximately linearly with the inlet gauge pressure. Similarly, at different critical states corresponding to various inlet gauge pressures, the relationship with the leakage rate is nearly linear in the ranges of 0–30 kPa and 60–300 kPa. However, between 30 and 60 kPa, the leakage rate exhibits a sudden increase. This is actually a symmetrical relationship around an inlet gauge pressure of approximately 48 kPa. Below 48 kPa, there is a sharp decrease in the pumping rate, while above 48 kPa, there is a sharp increase in the leakage rate, with both sides having nearly the same magnitude. At an inlet gauge pressure of approximately 48 kPa, the leakage rate is almost zero, and the critical rotational speed is approximately 14 r/min. This pressure can be referred to as "boundary-line pressure," similar to boundary-line speed, indicating the point where the leakage direction changes. It is the boundary-line at which the leakage rate transitions between leakage and pumping when other parameters remain constant. In seal design, attention must be paid to the boundary-line pressure to optimise performance and prevent undesired leakage or pumping effects.

From Equation (1), it is evident that as the inlet gauge pressure increases, the closing force increases linearly, and the corresponding film opening force also increases linearly. Therefore, the critical rotational speed follows a similar upward trend, as a higher rotational speed is required to generate sufficient film opening force to balance the forces and reach the critical opening state. Additionally, the increase in leakage rate can be explained by the greater pressure difference between the inside and outside of the seal system, which enhances the driving force for the two-phase fluid to flow through the leakage channel.

3.4.3. Influence of Temperature on Start-Up Performance

Taking the data from both Tables 1 and 2 for an inlet with a gauge pressure of 100 kPa, an oil–gas ratio of 0.03, and temperatures incrementally increasing from 40 °C to 130 °C, the variations in critical rotational speed and leakage rate were calculated. The results are presented in Figure 13.

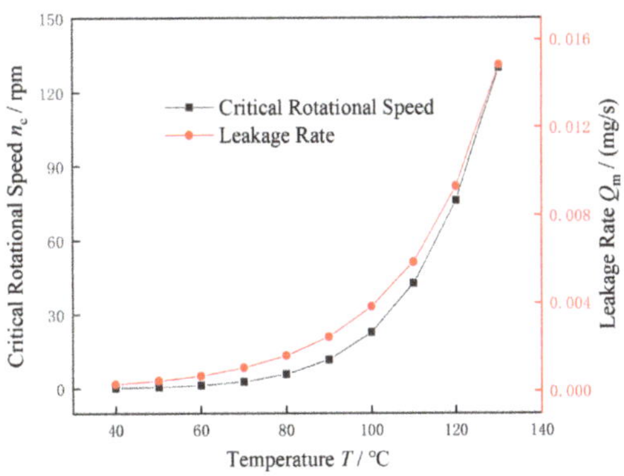

Figure 13. Critical rotational speed and leakage rate at different temperatures.

Figure 13 demonstrates that temperature has a significant impact on the start-up performance of the sealing system. Both the critical rotational speed and the leakage rate show a clear upward trend as the temperature increases. In the lower temperature range, such as from 40 °C to 80 °C, the changes in critical rotational speed and leakage rate are relatively gradual. However, as the temperature continues to rise, particularly beyond 100 °C, the increases in both parameters become more pronounced, with the curves steepening. This behaviour can be attributed to higher temperatures reducing the equivalent viscosity of the oil–gas mixture, thereby enhancing its fluidity and significantly increasing the leakage rate. Additionally, higher temperatures decrease the stiffness of the oil–gas two-phase film, reducing its load-bearing capacity. As a result, higher rotational speeds are required to reach the critical state for the seal.

In the cooling systems of electric spindles in new energy vehicles, the coolant temperature is typically controlled within the range of 40 °C to 70 °C. Some high-performance electric spindles may allow the coolant temperature to approach 80 °C, but this usually requires more efficient cooling designs to ensure that critical components, such as bearings and motor stators, are not adversely affected. In contrast, the temperature range for traditional automotive cooling systems is generally between 75 °C and 105 °C. This suggests that this new type of seal is more suitable for new energy vehicles than for traditional vehicles. Moreover, in the lower temperature range, the corresponding critical rotational speed is lower, making the start-up process of this seal in new energy vehicles easier, shorter, and with less impact from rubbing during start-up.

4. Conclusions

Through in-depth analyses of the start-up process for high-speed micro-grooved pumping seals in new energy vehicles, this study revealed the impact of various operating parameters on sealing performance.

Firstly, when the film thickness was below the critical value, the film opening force was calculated as the difference between the closing force and the surface contact force. Changes in the fractal dimension and characteristic length scale influenced the initial film thickness. When the film thickness exceeded the critical value, the balance between the film opening force and the closing force caused the film thickness to increase non-linearly with rotational speed, reflecting a decrease in the stiffness of the oil–gas two-phase film. This further highlighted the trend of leakage rate changes at high rotational speeds. Notably, the study found that under the given operating conditions, as the rotational speed increased, the sealing system transitioned from a leakage state to a pumping state, with a boundary-line speed of approximately 102 r/min. This finding provided important operational references for the practical application of sealing systems.

Additionally, operating parameters such as oil–gas ratio, inlet gauge pressure, and temperature significantly affected the critical state. Changes in the oil–gas ratio showed a non-linear decrease in critical rotational speed, suggesting that the design process should avoid an oil–gas ratio near 0.06 to reduce leakage rates during start-up. Variations in inlet gauge pressure also affected the critical rotational speed, with a sharp increase in leakage rate and a change in leakage direction near the boundary-line pressure, which was 48 kPa under the given conditions. The influence of temperature was evident in the significant rise in both critical rotational speed and leakage rate, emphasising the importance of temperature control in the cooling systems of new energy vehicles.

In summary, this study provided a systematic theoretical foundation and guidance for optimising the start-up process of high-speed micro-grooved pumping seals in new energy vehicles. Future research could further explore the dynamic sealing performance after opening and the microscale porosity leakage in the stationary state before start-up, with the goal of achieving more efficient seal designs.

Author Contributions: Conceptualisation: H.C., X.D. and R.Y.; methodology: H.C., X.D. and R.Y.; software: H.C. and R.Y.; validation: H.C.; formal analysis: H.C.; investigation: X.H. and X.B.; resources: X.D., R.Y., X.H. and X.B.; data curation: H.C.; writing—original draft preparation: H.C.; writing—review and editing: H.C.; visualisation: H.C.; supervision: X.D. and R.Y.; project administration: H.C., X.D. and R.Y.; funding acquisition: X.D. All authors have read and agreed to the published version of the manuscript.

Funding: This work was supported by the National Key Research and Development Program of China, grant number 2020YFB2010001.

Data Availability Statement: Data are contained within this article.

Conflicts of Interest: Authors Xianzhi Hong, Xin Bao were employed by Chengdu Yitong Seal Co., Ltd. The remaining authors declare that the research was conducted in the absence of any commercial or financial relationships that could be construed as a potential conflict of interest.

References

1. Jeremias, G.; Simon, F.; Frank, B. Wear on radial lip seals: A numerical study of the influence on the sealing mechanism. *Wear* **2021**, *476*, 203674.
2. Wen, C.Y.; Yang, A.S.; Huang, F.J.; Chang, H.T. New deflected-helix ribbed lip seal with enhanced sealing performance. *Tribol. Int.* **2011**, *44*, 2067–2073. [CrossRef]
3. Choi, H.J.; Park, C.W.; Lee, J.C.; Kim, J.G.; Choi, S.D. Analysis on the mechanical characteristics of PTFE oil seal for the rear part in the automotive engine. *Int. J. Precis. Eng. Manuf.* **2011**, *12*, 485–490. [CrossRef]
4. Huang, T.C.; Lin, C.Y.; Liao, K.C. Sealing performance assessments of PTFE rotary lip seals based on the elasto-hydrodynamic analysis with the modified archard wear model. *Tribol. Int.* **2022**, *176*, 107917. [CrossRef]
5. Mei, B.F.; Hu, B. Development and experimental research of a new type of low friction engine oil seal. *China Auto.* **2022**, *5*, 44–48.
6. Zhmud, B.; Najjari, M.; Brodmann, B. The effects of the lubricant properties and surface finish characteristics on the tribology of high-speed gears for EV transmissions. *Lubricants* **2024**, *12*, 112. [CrossRef]
7. Greenwood, J.A.; Williamson, J.B.P. Contact of nominally flat surfaces. *Proc. R. Soc. Lond.* **1966**, *295*, 300–319.
8. Mandelbrot, B.B. Stochastic models for the Earth's relief, the shape and the fractal dimension of the coastlines, and the number-area rule for islands. *Proc. Natl. Acad. Sci. USA* **1975**, *72*, 3825–3828. [CrossRef]
9. Johnson, K.L. *Contact Mechanics*; American Society of Mechanical Engineers: New York, NY, USA, 1985.
10. Majumdar, A.A.; Bhushan, B. Role of fractal geometry in roughness characterization and contact mechanics of surfaces. *Trans. ASME J. Tribol.* **1990**, *112*, 205–216. [CrossRef]
11. Majumdar, A.; Bhushan, B. Fractal model of elastic-plastic contact between rough surfaces. *J. Tribol. Trans. ASME* **1991**, *113*, 1–11. [CrossRef]
12. Wang, S.; Komvopoulos, K. A fractal theory of the interfacial temperature distribution in the slow sliding regime: Part I—Elastic contact and heat transfer analysis. *J. Tribol.* **1994**, *116*, 812–822. [CrossRef]
13. Dong, L.; Zhang, Y.X. Revision of M-B elastoplastic contact model. *J. Sichuan Univ. Sci. Technol.* **2001**, *2*, 4–7.
14. Ge, S.R.; Zhu, H. *Fractal of Tribology*; China Machine Press: Beijing, China, 2005.
15. Bhushan, B. The real area of contact in polymeric magnetic media—I: Critical assessment of experimental techniques. *ASLE Trans.* **2008**, *28*, 75–86. [CrossRef]
16. Bhushan, B. The real area of contact in polymeric magnetic media—II: Experimental data and analysis. *ASLE Trans.* **2008**, *28*, 181–197. [CrossRef]
17. Wei, L.; Liu, Q.H.; Zhang, P.G. Sliding friction surface contact mechanics model based on fractal theory. *J. Mech. Eng.* **2012**, *48*, 106–113. [CrossRef]
18. Ding, X.X.; Yan, R.Q.; Jia, Y.L. Construction and analysis of fractal contact mechanics model for rough surface based on base length. *Tribology* **2014**, *34*, 341–347.
19. Zhao, Y.X.; Ding, X.X.; Wang, S.P. Prediction of leakage rate and film thickness of mechanical seal based on fractal contact theory. *Lubr. Eng.* **2022**, *47*, 156–163.
20. Fukui, S.; Kaneko, R. Analysis of ultra-thin gas film lubrication based on linearized Boltzmann equation: First report—Derivation of a generalized lubrication equation including thermal creep flow. *Trans. ASME J. Tribol.* **1988**, *110*, 253. [CrossRef]
21. Gu, Y.Q. Just-lift-off phenomena and lift-off characteristics of mechanical face seals. *Petro-Chem. Equip.* **2003**, *4*, 26–29.
22. Li, S.X.; Song, B.W.; Zhang, Q.X.; Cai, J.N.; Gao, J.J. Opening characteristics of dry gas seal. *CIESC J.* **2011**, *62*, 7.
23. Fan, Y.; Song, P.Y.; Xu, H.J. Study on startup operation of dry gas seal with steam lubrication. *CIESC J.* **2020**, *71*, 3671–3680.
24. Sun, X.J.; Song, P.Y.; Mao, W.Y.; Deng, Q.G.; Xu, H.J.; Chen, W. Dynamic contact analysis of dry gas seal during start-stop process considering material properties and surface topography of seal rings. *CIESC J.* **2021**, *72*, 4279–4291.
25. Chen, H.Q.; Yan, R.Q.; Hong, X.Z.; Bao, X.; Ding, X.X. Micro-groove optimisation of high-speed inner ring micro-grooved pumping seal for new energy electric vehicles. *Processes* **2024**, *12*, 1281. [CrossRef]
26. *JB/T 11289-2012*; Specification for Dry Gas Seal. China Machine Industry Press: Beijing, China, 2012.
27. Yan, W.; Komvopoulos, K. Contact analysis of elastic-plastic fractal surfaces. *J. Appl. Phys.* **1998**, *84*, 3617–3624. [CrossRef]
28. Yan, R.Q.; Chen, H.Q.; Zhang, W.Z.; Hong, X.Z.; Bao, X.; Ding, X.X. Calculation and verification of flow field in supercritical carbon dioxide dry gas seal based on turbulent adiabatic flow model. *Tribol. Int.* **2021**, *165*, 107275. [CrossRef]
29. Wen, S.Z.; Huang, P.; Tian, Y.; Ma, L.R. *Principles of Tribology*; Tsinghua University Press: Beijing, China, 2012.

lubricants

MDPI

Article

In Situ Operando Indicator of Dry Friction Squeal

Maël Thévenot [1,*], Jean-François Brunel [1], Florent Brunel [1], Maxence Bigerelle [2], Merten Stender [3], Norbert Hoffmann [4] and Philippe Dufrénoy [1]

[1] Laboratoire de Mécanique Multiphysique Multiéchelle, LaMcube, UMR 9013, Centrale Lille, CNRS, University Lille, 59000 Lille, France; jean-francois.brunel@polytech-lille.fr (J.-F.B.)
[2] LAMIH-Laboratoire d'Automatique de Mécanique et d'Informatique Industrielles et Humaines, UMR 8201, CNRS, University Polytechnique Hauts-de-France, 59313 Valenciennes, France
[3] Mechanical Engineering Group, Berlin University of Technology, 10623 Berlin, Germany; merten.stender@tu-berlin.de
[4] Dynamics Group, Hamburg University of Technology, 21073 Hamburg, Germany; norbert.hoffmann@tuhh.de
* Correspondence: mael.thevenot@univ-lille.fr

Abstract: In various applications, dry friction could induce vibrations. A well-known example is frictional braking systems in ground transportation vehicles involving a sliding contact between a rotating and a stationary part. In such scenarios, the emission of high-intensity noise, commonly known as squeal, can present human health risks based on the noise's intensity, frequency, and occurrences. Despite the importance of squeal in the context of advancing urbanization, the parameters determining its occurrence remain uncertain due to the complexity of the involved phenomena. This study aims to identify a relevant operando indicator for predicting squeal occurrences. To this end, a pin-on-disc test rig was developed to replicate various contact conditions found in road profiles and investigate resulting squealing. Each test involves a multimodal instrumentation, complemented by surface observations. It is illustrated that the enhanced thermal indicator identified is relevant because it is sensitive to the thermomechanical and tribological phenomena involved in squealing.

Keywords: brake squeal; thermal solicitation; surface topography; friction-induced vibrations; tribology

1. Introduction

Braking function is a critical element in any vehicle. Whereas various braking technologies are in use, friction braking still remains the most prevalent. It is considered as the most effective and reliable, which is of particular interest for a component used in emergency situations. Particular attention has been paid over the years to the development of these braking systems, which realize this conversion of the vehicle's kinetic energy into other forms, primarily heat, as the brake pads interact with the brake disc. However, the use of friction brakes still involves significant unresolved problems. Thus, friction instabilities, thermal fatigue [1], wear [2] and particles or noise emissions such as squealing [3] remain as significant challenges to the improvement of these systems. The occurrence of brake squeal in vehicles is one of the most expensive issues in industrial applications (after-market) [4], and significant efforts are made to mitigate this [5]. It is a particularly problematic issue for people, due to its high frequency and acoustic pressure (above 1 kHz and starting at 70 dB). This challenge is all the more true for electric vehicles, which are even more sensitive to noise due to the absence of noise-covering combustion engine sounds. It is commonly accepted that squealing comes from self-excited vibrations of the brake system, which endure unstable dynamical behavior (like mode coupling) under sliding frictional contact conditions [6]. But a complete understanding of its physics is still lacking, as it remains a complex phenomenon affected by multiple factors on multiple scales from the behavior of structural components to external excitations, as well as the dynamics of the friction interface during braking.

Over the years, to enhance our understanding of the mechanisms leading to squeal events, researchers have conducted experiments and numerical studies to investigate the behavior of the contact interface. Experimental studies shown that within a sliding contact, the evolution of imposed conditions [7,8] and macroscopic aspects [9], such as changes in the friction material, surface's geometry and tribolayer (thin film at the interface) are interconnected and have an influence on the squeal occurrence [10]. Different rotation speeds of the rotating part, for example, lead to the evolution of vibration responses and contact behaviors. Relationships can be observed, such as a stronger vibration response occurring at lower disc rotation speed, like 200 rpm, leading to high sound pressure level [11] and also severe localized wear and rough pad surface. Moreover, research works reported a strong correlation between wear and squeal occurrence in braking systems [12]. From prepared parts [13], successive contacts gradually modify the contact behavior with an elevation of the friction coefficient until it starts to plateau around a critical value. Squeal occurrence sharply increases once this contact behavior is reached, coinciding with increased use of parts. However, it has been observed that the friction coefficient and its variation are not enough to explain squeal occurrence, as many studies address the fact that mode coupling can take place under a constant friction coefficient. This coefficient should rather be viewed as a response of the tribological system. Finite element models employing complex eigenvalue analysis (CEA) have also demonstrated the sensitivity of components (dynamic modes) and system parameters (assembly with connections) on squeal propensity [14]. The influence of thermomechanics and friction materials on the dynamic response has also been highlighted [15,16]. However, the models are still not predictive.

Additional studies focused on the evolution of friction material and tribology [17], demonstrating the complexity of involved mechanisms, as the friction material undergoes modifications due to thermal, mechanical and tribological loading as far as the tribolayer. On both mesoscopic (contact geometry) and microscopic (particles flow) scales, it is known from the third-body approach proposed by Godet [18] and Berthier [19] that the flows and compaction of worn particles (which constitute the major part of what is called the third body) play a crucial role in the creation of the tribological surface layers [20]. These changes at a microscopic scale such as evolution of the contact surface heterogeneity (due to material degradation and presence of a third body) significantly impact squeal occurrence. This is interconnected with the tribological circuit and its complex behavior [21]. It has been shown that without changes to the interface (no wear of friction material, and hence no third body in the contact and constant macro contact surface), a tribological system remains silent and squeal only occurs upon the addition of a third body. Squeal occurrence is then highly dependent of the size and nature of particles flowing through the contact [22], in addition to the rest of the contact conditions (friction coefficient, dissipated energy, etc.). Numerical works also show that the emergence of squeal is dependent on the evolution of the surface and material transformation beneath the surface of friction material [23].

Most recent experimental studies addressing squeal issues and tribology in general tend to include the observation of the worn surface after contact [24]. The added value of these observations is, however, often limited, as after opening the contact, a significant amount of critical information is lost, particularly if the observation requires a lot of manipulation of the parts. The distribution of contact plateaus in the tribolayer, which are formed by hard ingredients and wear debris compaction around them, have been shown to have an influence on squealing events [25]. The impact of the changes in the actual contact localization at macro and meso scales on the squealing frequencies of a pin-on-disc system have also been highlighted [26]. This research work emphasizes how the evolution of the contact state during dry friction significantly influences the squeal occurrence. However, the contact is not accessible during contact, which makes characterizing its state only feasible before and after the tests. Therefore, it would be very useful to have an indicator to identify the conditions in which squeals emerge, in order to anticipate how to avoid them. In this work, we propose to explore indicators other than global ones, such as friction coefficient or disc temperature, which are often linked, but only partially correlated with squeal noise. To this end, the indicator must be related to the phenomena described above, hence the idea of a carefully chosen intrusive measurement. Thermomechanical constraints on the friction material are monitored during tests with embedded thermocouples and Foucault's displacement sensors, while a microphone records the noise. A discrete tracking of the friction material surface with a remote head profilometer is introduced, allowing for optical observations as well as profilometry without disassembling the parts.

2. Methods

2.1. Experimental Setup

For this study, a pin-on-disc (PoD) tribometer developed for another study [9] was adapted to perform the required test conditions (Figure 1). The interest of this set-up is to master the dynamic behavior of the system by the limited assembly. It is made of a rigid stand on which an aluminum blade is attached to its extremities. The pin is glued at the center of the blade. The imposed displacement of the rigid frame with a motorized table along the axial disc direction pushes the pin against the disc and the bending of the blade gives the normal load, as detailed in Section 2.3.

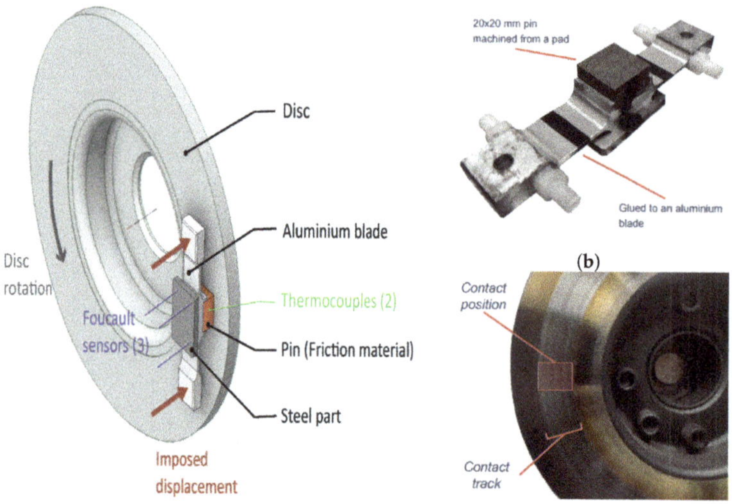

(a) (c)

Figure 1. (**a**) Schematic representation of parts in contact. (**b**) Stationary part: pin of friction material mounted on aluminum blade. (**c**) Rotating part: steel disc with contact location.

The 20 × 20 mm pin is machined from a complete pad, with a height of 10 mm. A diamond disc polisher is used to ensure the dimensions of the pin. The disc has an external diameter of ∅220 mm and the mean contact radius on the disc is set at 100 mm, as illustrated in Figure 1. Materials used for this study are detailed in Section 2.4.

2.2. Testing Conditions

The test campaign is divided into series of 20 successive friction tests with fixed conditions, including contact and non-contact conditions. Friction tests are performed with a constant disc rotation speed and an imposed displacement to obtain the desired normal load at the start of the test. The rigid frame position is then kept constant during all the test series, leading to a variation of the normal load due to thermomechanical solicitations (thermal dilatation and wear).

Different test configurations can be achieved by modifying the desired normal load (200 N, 400 N), the disc rotation (200 rpm, 600 rpm, 700 rpm), the contact duration (10 s,

30 s) or the non-contact duration between successive tests (5 s, 10 s, 30 s). The reduction in the non-contact duration leads to a higher temperature elevation of the disc and the pin. This approach allows us to perform a large number of tests; for this test campaign a reduced combination of solicitations have been explored, which allowed us to repeat them a large number of times.

Figure 2 presents the test conditions investigated through the whole campaign. Disc rotation speed is represented, as well as the pin's axial starting position through the campaign. Its position is updated when necessary to ensure that a fixed displacement led to the desired normal load at the beginning of each test series (200/300 N). The global evolution of the pin's axial position through the campaign demonstrates the wear of the pin. Duration during (ON) and between (OFF) each of the tests is also indicated. Finally, between each test series, a red dot on the line of the disc's rotating speed corresponds to instants where an optical observation of the pin's surface is performed.

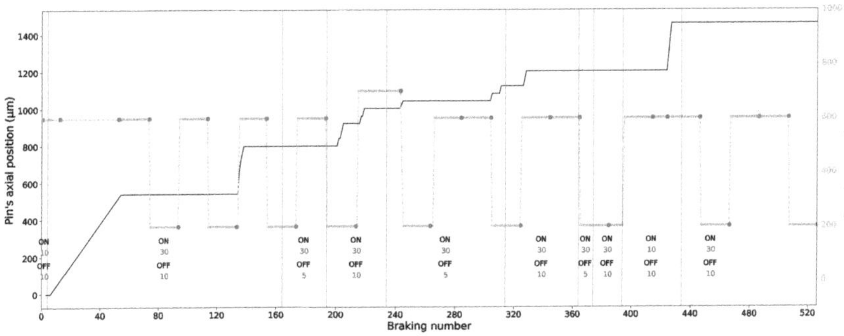

Figure 2. Test conditions investigated in this study in terms of rotation speed of the disc and pin's axial displacement. Duration of test (ON) and between each test (OFF) are indicated in blue in seconds and pin's surface observations with red dots.

2.3. Instrumentation

A dedicated instrumentation has been implemented to characterize thermomechanical constraints during the contact process.

Three Foucault displacement sensors measure the aluminum blade's bending. Knowing the blade's stiffness, from those sensors, the contact normal load is obtained. This relationship between displacement and force is obtained by prior calibration. Two type K thermocouples with a diameter of ∅1 mm are inserted in the pin, located 2 mm below its surface. Both thermocouples are in the same radial position (regarding the disc), one toward the leading edge (TC1) and the other toward the trailing edge of the contact (TC2), as illustrated in Figure 3. Sound from the tests is recorded using a microphone, the range of which is from 3 Hz to 20 kHz. It is located at a distance of 1 m from the disc. Mechanical sensors and microphone signals are recorded using an high rate acquisition sampling rate of 50 kHz, while thermocouples are recorded at a sampling rate of 90 Hz.

Lastly, a profilometer (Alicona, PortableRL model) is integrated into the test bench to conduct discrete tracking of the pin's surface. Based on the technology of focus variation, it can provide information regarding topography. Its white-light lighting allows fixed observation conditions through a test campaign, to be able to compare images taken with fixed settings. For this purpose, the steel frame holding the pin's blade can be rotated to two predefined target positions: either in a test configuration to perform contacts (pin in front of disc) or in an observation configuration (pin in front of profilometer). Mechanical clamping for these target positions ensures that they remain the same throughout the campaign. Figure 3 illustrates the pin's placement in both configurations, with indications of the thermocouple positions. The pin's surface is examined through a 4× objective with a vertical resolution of 1.4 µm and a lateral resolution of 7 µm, using stitching techniques to enable an image of the whole pin surface to be obtained (25 stitches to cover the 20 × 20 mm surface).

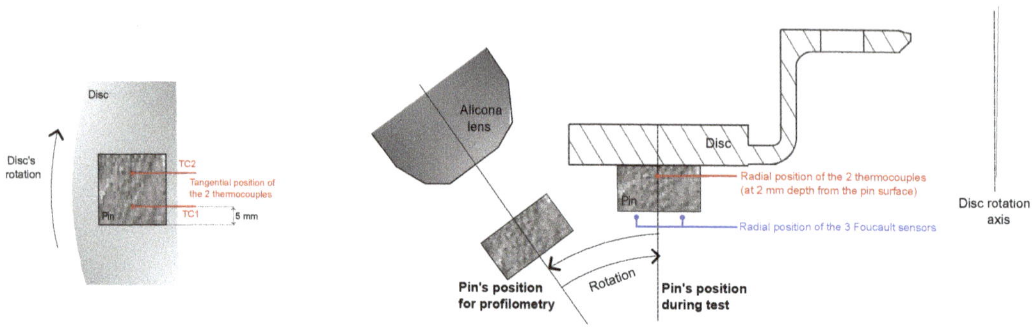

(a) (b)

Figure 3. (**a**) Top view of the test bench presenting the pin's positions for test and observation. (**b**) Front view of the pin in testing position.

2.4. Materials

A disc made of steel and a commercial low-met friction pad material designed for railway applications are used. As this material is used in industry, its formulation is only partially known, without details about raw components or process (except that the maximum processing temperature is 200 °C, and that the surface undergoes grinding without subsequent polishing). SEM observations have been conducted of the material before tests to have a better understanding of its composition. Some components have been identified either through EDX analysis or due to their particular shape. The material demonstrates a matrix made of resin and rubber, with metallic fibers. The friction surface is taken as if it is on pads, i.e., after machining by material removal.

An optical observation of the pin's surface is realized before the test campaign with the profilometer, as shown in Figure 4a. At this macroscopic scale, the material presents a high degree of heterogeneity. Patterns of different colors with a size around 3 to 4 mm are observed. Some are darker and others are brighter with an orange tint.

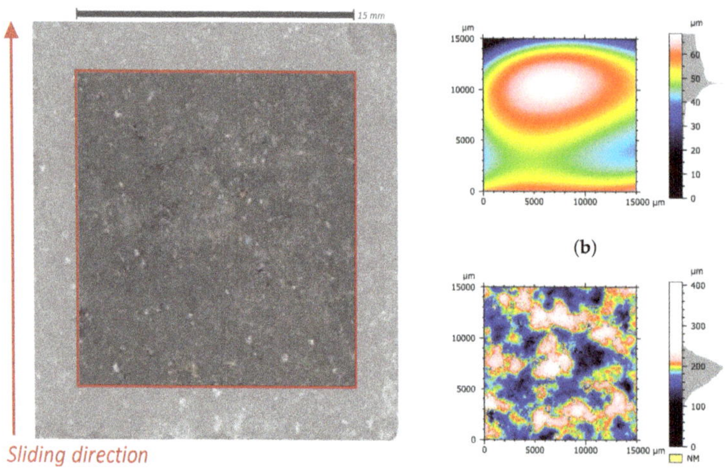

(a) (b) (c)

Figure 4. Profilometry observation of the pin before tests. (**a**) Image of the 20 mm × 20 mm pin's surface, with some brighter patterns highlighted. (**b**) Shape of the pin's surface. (**c**) Pin's profilometry without the shape.

In addition to the optical observation, the profilometer allows us to obtain the topography of the pin's surface. To correctly investigate the pin's profilometry, a pre-processing of the data is carried out. First, aberrant or unmeasured points are removed. A windowing operation is then performed to retain only a region of 15 mm by 15 mm centered at the center of the pin (highlighted center area in Figure 4a). This allows us to remove the edges from further operations. A straightening operation is then carried out by removing a least squares plane from the surface. Finally, the shape is removed from the image with a third-degree polynomial (Figure 4b). The resulting surface (Figure 4c) contains the waviness and roughness, without the global shape of the pin. In this surface, patterns with the same scale as those noticed on the image are visible with a difference in terms of height compared to the rest of the surface. It confirms a strong heterogeneity at a millimeter scale (around 3–4 mm), probably due to the formulation and processing of the material.

3. Results

3.1. Global Results

Figure 5 presents the thermal and mechanical measurements along the campaign. For each test, the maximum normal load and the maximum temperature for the two thermocouples are shown. Disc rotation speed is still represented to facilitate the comparison with Figure 2. Each test series is visible on the pin temperatures, which rise during the series. Depending on the imposed normal load, contact duration and rotating speed of the disc, the temperature rises are different during the cycle. Overall, the thermocouple at the trailing edge (TC2) displays higher temperatures than the thermocouple at the leading edge (TC1) of the contact. It should be noted that the TC2 was detached during one series (from test 425 to 445). For the majority of the test campaign, maximal temperatures are correlated with maximal normal load elevation, with the exception of the first temperature rise above 200 °C, where the normal load decreased during temperature elevation. Enduring high thermal solicitations, pin materials may have undergone transformations.

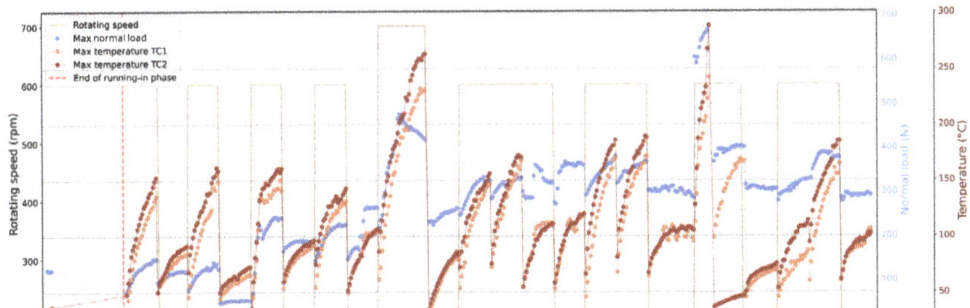

Figure 5. Thermomechanical conditions investigated, resulting from the imposed test conditions.

3.2. Surface Analysis

Besides the continuous measurements of thermomechanical constraints performed during tests, discrete tracking of the pin surface is achieved through optical and profilometry measurements. Those observations allow us to track the evolution of the surface state before and after each cycle of tests. Comparing the images allows us to track some material flows, such as the appearance or ejection of particles, as well as changes in the powder deposited on the surface, whereas profilometry gives an evolution of the topography of the pin surface through the campaign. Profilometry observation before tests showed a pattern on the pin's surface, with significant amplitude variations every 3–4 mm. In order to track their evolution through the campaign, it is necessary to have a quantitative indicator of these patterns. Surfaces from which shape is removed, thanks to the previously presented post-processing, allow us to calculate a range of surface roughness parameters to characterize certain aspects of the surface. First, a certain cut-off needs to be chosen to determine at which scale the surface is characterized, then roughness parameters can be calculated for each surface. A statistical analysis (MESRUG) was performed to compare the relevance of roughness parameters regarding the surface state [27]. Among the various calculated roughness parameters, S5p with a specific cut-off of 4413 µm using a non Gaussian filter emerged as the most representative parameter for capturing the evolution of observed patterns. Figure 6 represents the S5p of each profilometry observation of the campaign.

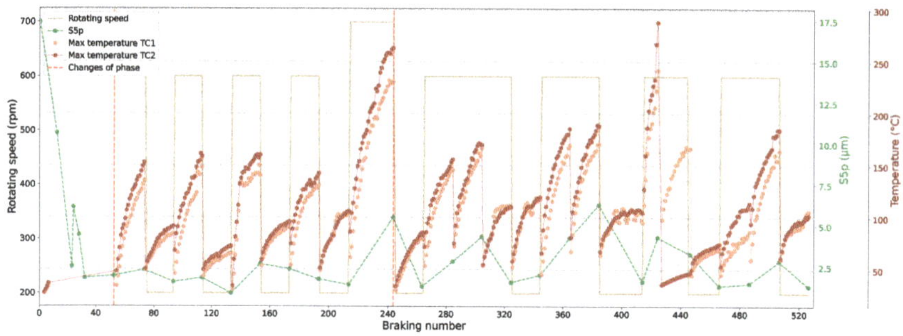

Figure 6. Evolution of the surface roughness parameter S5p during the test campaign.

There is a quick decrease in the S5p during the first 40–50 tests of the campaign. It is related to the running-in phase at the beginning of the test campaign, during which there is a rise in the macroscopic load-bearing area between the parts until the contact occurs across the entire pin surface. After 50 tests, the optical observation showed that the contact front went through the entire pin surface, and the running-in phase is considered complete. The characterization of the surface through this roughness parameter highlights the change in amplitude on the surface during the running-in phase, due to the erosion of the initial peaks. The S5p stabilizes with the evolution of test series conditions; most energetic tests induce an increase in the parameter, and vice versa. During the first 210 tests, the S5p at this scale does not show much variation. Then, the cycle with the first rise in temperature above 200 °C that the pin endures is realized. During this cycle, an important variation of the S5p is observed. For the rest of the campaign, the S5p continues to show a lot more variation when the pin's temperature rises. During these tests, the material roughness at the observed scale provides some insights into past thermal conditions.

Regarding the pattern amplitude on the topographies, it appears that some surfaces present a lot of similarities to each other, in correlation with the S5p parameter. To investigate phenomena and solicitations responsible for those changes in the surface, topographies were grouped together in classes according to the similarity of the pattern. Four distinct classes could be established (Figure 7). Class A is characteristic of the pin's state before tests and is never observed again during the campaign (high value of S5p). During the rest of the campaign, the three other classes appear one after another regarding the test conditions (Figure 8; two images are marked "unmeasurable", as the measured images were incomplete). In general, higher temperature levels lead to surfaces that fall in class B (medium value of S5p), while lower constraints lead to surfaces in class D (with class C in between, with a low value of S5p). To have a better understanding of those resulting topographies, images of the surfaces are shown in Figure 7 as well. On the surfaces, some powdery matter is observed (dark color in asperities on the images). It is identified as a powdery third body that flows through the contact during the test, and can be trapped in asperities and porosities of the surface, and sometimes accumulate enough to participate in the contact, showing compaction. For all images, the more powdery matter is present on the surface, the more the surface falls into class D (surfaces A and B showing less powder).

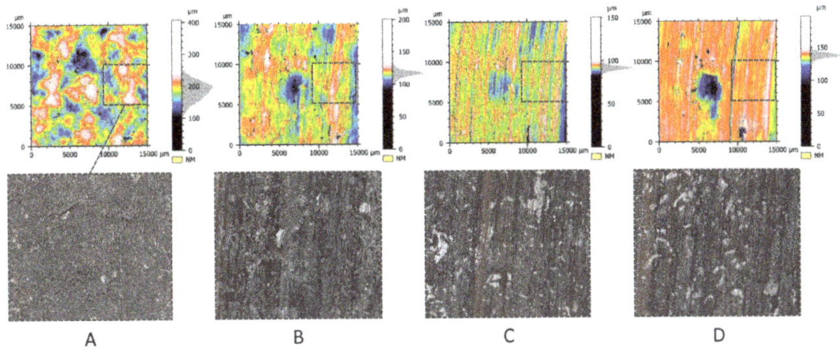

Figure 7. Classes of surface topography observed with a zoomed area of each corresponding image: (**A**) very rough aspect, (**B**) rough with a noticeable friction direction, (**C**) clear sliding direction and smaller high differences, (**D**) rather smooth surface with a local porosity.

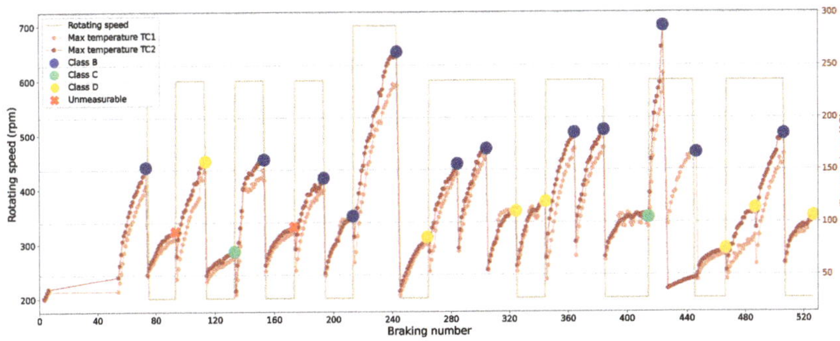

Figure 8. Classification of the profilometry observations through the test campaign.

Surface observations highlighted a link between tribological circuit and solicitations. Contact conditions inducing lower thermal solicitations on the parts increases the storage capacity of powder within the interface. The contact area appears more enclosed when thermal phenomena are limited. This inclination to favor powder storage at the interface leads to the filling of surface irregularities, resulting in a smoothing effect (and a lower S5p). Higher thermal solicitations result in rougher surfaces and emptier asperities. Intense thermal phenomena lead to more pronounced opening of the interface and facilitate particle ejection. Squeal phenomena are associated with variations in the effective contact area between components. The impact of these surface state variations under different thermal solicitations on squeal occurrence will now be investigated.

3.3. Noise Emission Analysis

In this paper, noise emissions are considered as squeal when the threshold of 70 dB is reached for any frequency. The frequency range of concern spans from 1000 Hz (excluding bench noise) to 20 kHz. Figure 9 provides an overview of the campaign with the squealing frequency of highest intensity and its harmonics for each test. To process the signal from the microphone, which has a 50 kHz sampling frequency, a Short-Time Fourier Transform is applied to obtain the acoustic spectrogram.

The disc's rotation speed is presented as well. Different frequencies are obtained with the presence of harmonics which are characteristic of squeal. As seen before with the surfaces, different phases of the campaign result in changes in the noise emission. Most of the silent tests occur at the beginning of the campaign, where the normal load is low and the contact area evolves a lot during each test or during the beginning of each test series.

During the first phase of the campaign up to brake number 210, only one squeal frequency at 2.1 kHz (with its harmonics) is obtained, which appears to be sensitive to the test conditions. During the high-temperature series, high thermal solicitations lead to a new squeal frequency (5.3 kHz). This squeal frequency is then obtained for each test series where the temperature is high. At low temperature, two other frequencies are observed (3.3 kHz and 18 kHz). The three first low frequencies are very close to the disc's natural frequency [26]. It should be noted that the frequency at 2.1 kHz no longer appears after the first high-temperature test series, which may be related to a strong transformation of the contact interface and certainly of the friction material near the surface.

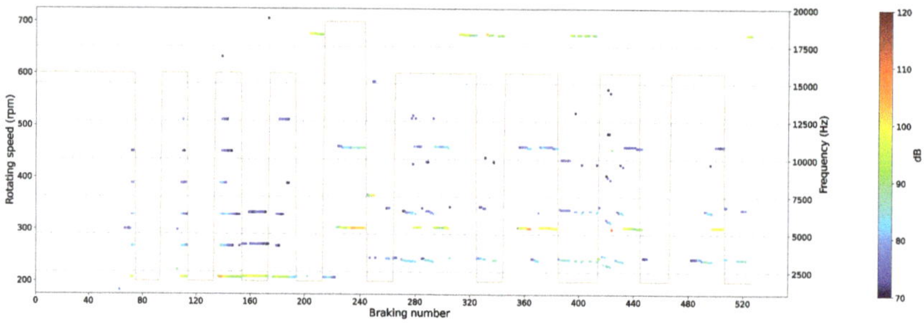

Figure 9. Overall acoustic results during the campaign: squeal detection for each frequency > 70 dB, disc's rotating speed and maximal temperature of the pin.

To highlight the main squeals of this campaign of more than 500 tests, a classification of the different obtainable spectrograms has been performed. Similar to what was performed for the surface, the idea is to identify the mechanisms responsible for the main behavior that emerges. For the whole test campaign, the harmonics are removed and a +/−500 Hz window gathers close squealing frequencies together. It results in only four different fundamental squealing frequencies. Each of these frequencies was associated with a squealing class for the classification (plus a class corresponding to an absence of squeal). Figure 10 presents some examples of spectrograms obtained during the individual test, associated with each identified squeal class. This classification is rather simple, so the duration of the squeal is not taken into account (as long as the frequency is above 70 dB for at least 1 s, it is classified in the corresponding category), and if multiple of those squealing frequencies appear in one test, it is added to all the corresponding classes.

Figure 11 resumes the spectrogram classes through the campaign. It can be noted that some tests fall in multiple classes. It is observed that classes No Squeal and 2.1 kHz Squeal are mostly observed before the transformation cycle occurring at test 215, and at rather low solicitations for class No Squeal. For the rest of the campaign, the previously observed correlation with the maximal temperature of the pin appear even more clearly. Class 5.3 kHz Squeal is the only frequency observed when the pin's temperature is above 140 °C, whereas at lower temperatures, behaviors appear more complex. Multiple squeal frequencies share the same range of temperature as classes 3.3 kHz Squeal and 18 kHz Squeal are observed. So the level of temperature seems to be a strong indicator, especially for the triggered squeal frequency at high temperatures. On the other hand, temperature variations allow us to make the distinction between some squealing frequencies at low temperature levels. When the squealing class 18 kHz Squeal occurs, the temperature fluctuates around a constant level, and the temperature of the TC1 thermocouple is higher than the temperature of the TC2, contrary to the squealing class 3.3 kHz Squeal, which appears only during temperature rises, with maximum temperature at the TC2. So the temperature is, for this test setup, a discriminating indicator for squeal frequencies, and beyond the temperature level, the heat kinetics and distribution are also important factors.

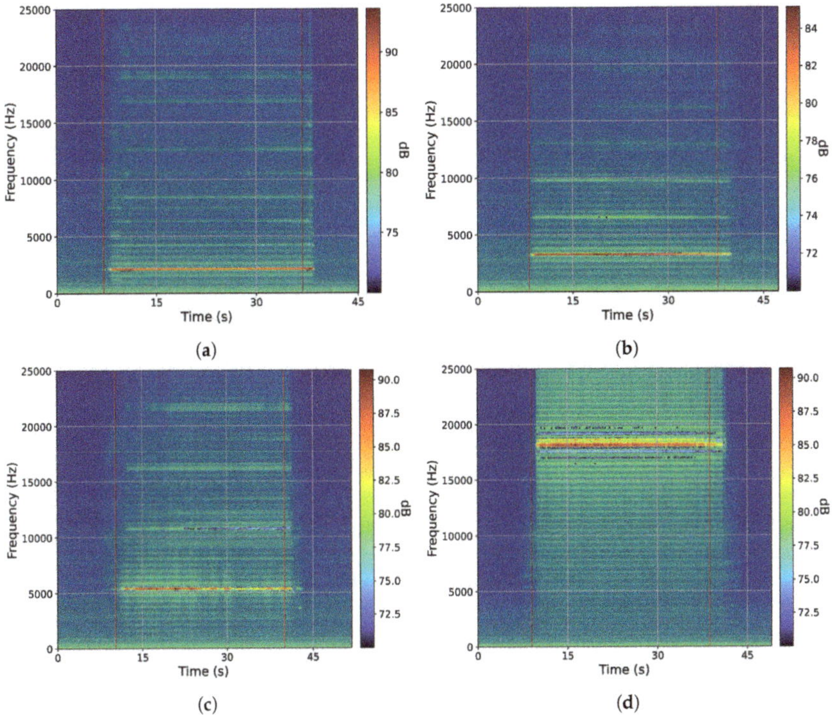

Figure 10. Classification of noise emissions for individual test. (**a**) 2.1 kHz squeal, (**b**) 3.3 kHz squeal, (**c**) 5.3 kHz squeal, (**d**) 18 kHz squeal.

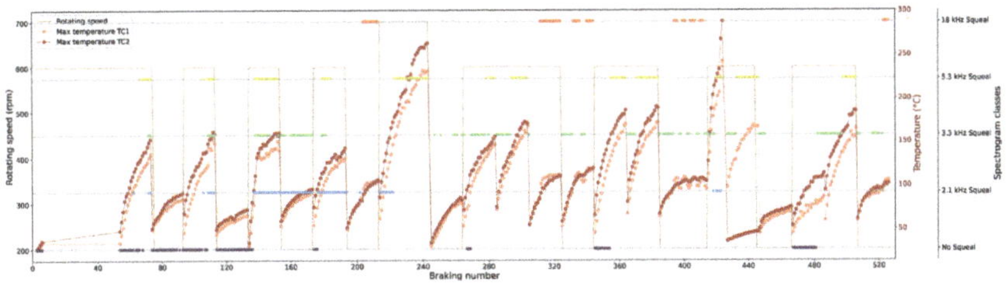

Figure 11. Classes of noise emissions observed for each test through the campaign, with rotating speed of the disc and pin temperature.

4. Discussion

Figure 12 presents an overview of all frequency classes, regarding the max temperature of the pin (maximum of the two thermocouples) in the last test performed just before the profilometry acquisition. The surface class is also indicated to compare with squeal classification. Class *A* for initial state has been removed here, and points circled in the gray tinted area correspond to observations realized before the transformation cycle with high thermal solicitation (test 0 to 215). As seen before, surfaces with topography of class *B* are only obtained if the pin temperature previously reached more than 140 °C. And those tests only correspond to squeal class *2.1 kHz Squeal* during the running-in phase, or *5.3 kHz Squeal* for the rest of the campaign. Below 140 °C, behavior appears more complex. Surface class *D* seems to occur most of the time for the lower temperature, but there is an overlap in term of temperature with the surface *C*, which is also present around 110/120 °C. A connection is highlighted between maximum pin temperature, squeal classes and surface classes.

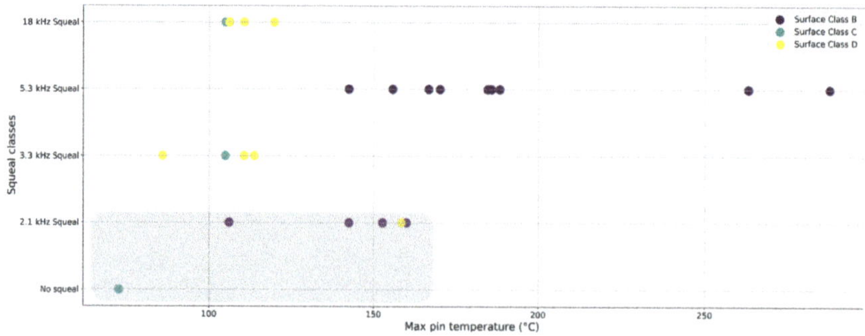

Figure 12. Representation of frequency classes regarding maximum temperature of the pin, with surface classes for corresponding profilometry observation.

Relationships have been established here between surface and squeal classes regarding the temperature solicitation, but only considering the maximum pin temperature of the test before profilometry. Yet, the profilometry observations are relatively scarce in comparison to the number of tests performed and the temperature level reached alone does not suffice to indicate squeal frequencies at lower temperatures. As mentioned before, temperature variations and heat kinetics on the surface area must also be considered to differentiate noise emissions. The impact of a test cycle reaching temperatures sufficient to potentially alter the material of the pin must also be acknowledged. The strong correlation between maximum temperatures and the pin surface state take places after this intense thermal solicitation.

A more in-depth analysis of the thermal solicitation on squeal for tests after the transformation cycle (tests 240 to 526) is presented in Figure 13. The maximum difference of temperature between the two thermocouples is displayed (trailing edge minus leading edge thermocouple) regarding the maximum reached temperature. All resulting points are connected to the centroïd (center of mass represented by a triangle) of their respective point clouds, which have the convex hull of their area colored for a better visualization. For this representation, squeal class *2.1 kHz Squeal* is excluded due to its limited number of data points. The number of displayed tests is still more substantial than the number of profilometry observations, allowing for the verification of previously observed hypotheses across a larger sample group.

As previously noted, the maximum pin temperature remains a distinguishing factor for class *5.3 kHz Squeal*, with only a few tests of class *3.3 kHz Squeal* falling within the same temperature range. Some of these few tests of class *3.3 kHz Squeal* showing a maximum pin temperature above 150 °C correspond to tests 417 to 419, which takes place in a high-temperature test series with potential material transformations being able to impact squeal frequency. The rest of these tests correspond to tests 496 and 497, where a squeal at 3.3 kHz occurs only at the beginning of the test and disappears when the temperature rises. So those points should in reality be linked with a lower temperature than the maximum pin temperature reached during the whole test. A more precise classification of squeal frequency regarding pin temperature would allow for an even more clear discrimination of temperature domains of each squealing frequency.

At lower temperatures, although classes *No Squeal*, *3.3 kHz Squeal* and *18 kHz Squeal* share a similar range of maximum pin temperature, clear differences emerge in terms of the temperature gap between the two thermocouples. Class *18 kHz Squeal* in particular is clearly delineated on this graph, appearing from 100 °C and only for low or negative temperature differences. While, in the majority of tests, the thermocouple at the trailing edge reaches higher temperatures, for some tests, the one at the leading edge of the contact exhibits higher temperature, triggering this particular frequency. This variation in temperature differences can be associated with an uneven distribution of contact pressure, which indicates a change in the load bearing. This change in bearing capacity may itself be linked to thermomechanical effects (e.g., dilatation) or interface variations (tribolayer or material detachment).

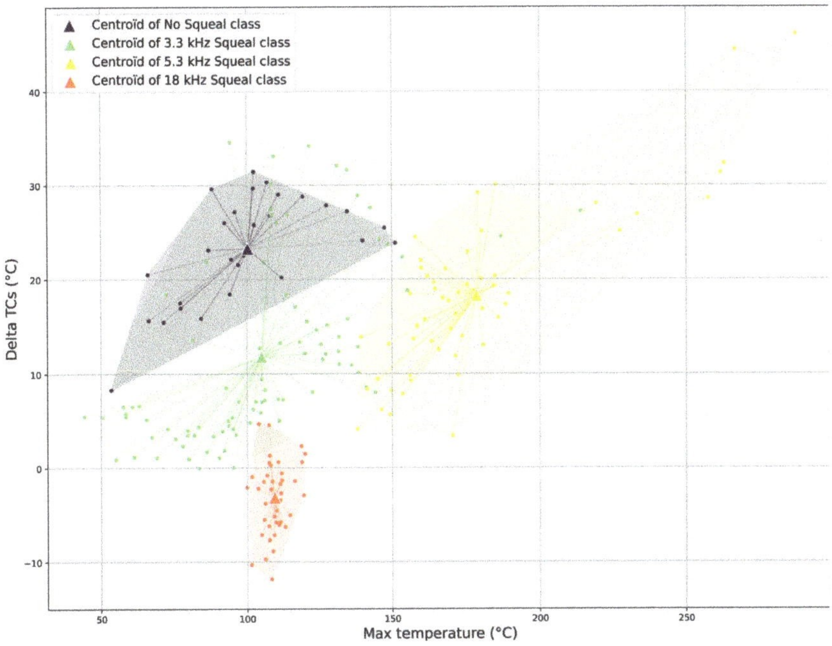

Figure 13. Temperature difference of pin thermocouples for tests 240 to 526, regarding the maximal pin temperature.

In contrast, the absence of squealing (class *No Squeal*) is noted between 50 °C and 150 °C, but only when the thermocouple at the trailing edge records significantly higher temperatures than its counterpart. Class *3.3 kHz Squeal* displays the widest range in both maximum pin temperature and temperature differences. However, in most tests where this frequency is triggered, the temperature rise at the trailing edge exceeds that at the leading edge by a margin of around 10 degrees. In this representation, classes *No Squeal* and *3.3 kHz Squeal* still overlap, yet their centroids suggest that a load-bearing area predominantly located on one side of the surface (specifically, at the trailing edge of the contact) tends to favor the absence of squealing. In this configuration, the exit of the contact is more confined, and the internal flow of particles through the contact might be limited if they are mostly emitted near the exit of the contact.

The present study has highlighted a significant correlation between pin temperature and squeal occurrence during dry sliding contacts. This finding aligns with previous research emphasizing the crucial role of temperature in tribological phenomena associated with these events. Investigation of the pin's profilometry has shown a robust correlation between the maximum pin temperature and the resulting surface condition. This suggests that the temperature seen by the material is a main contributing factor to the surface state of the pin. Through the imagery of the profilometry, it appears that the behavior of the third body through the contact and its ability to resorb asperities of the surface, hence lowering its macroscopic roughness, is significantly impacted by the temperature. These results have been confirmed by other experiments on the same device. An extension to other systems would require further investigation, similar to those proposed here, i.e., extended instrumentation for temperature measurement to obtain information on contact localizations. These localizations depend on the rigidity of the system, as does the vibratory response.

Lower-temperatures solicitations allow for a more powdery third body to be trapped in asperities, reducing surface roughness parameters on a scale of several millimeters. It might result in more opportunities for the load-bearing area to evolve as powdery material accumulates or is compacted, thereby triggering a greater variety of squealing frequencies. Conversely, higher thermal solicitations lead to the clearing of asperities and porosities, amplifying the role of millimetric patterns on the surface in contact localization and resulting in more specific and stable squealing frequency. Monitoring the pin's temperature during tests could potentially allow some squealing frequencies to be entirely avoided.

5. Conclusions

The complexity of the phenomena involved in squeal noise explains why, despite the huge amount of work dedicated to this issue, predicting them remains a challenge. It was proposed in this work to identify whether a physical indicator could be obtained to predict squeal situations. It was shown that an operando thermal measurement could enable this prediction. The relevance of this measurement is explained by the links with the physical phenomena influencing squeal, i.e., macroscopic and microscopic contact in relation to thermomechanics and tribology. It appears that variations in these phenomena can be captured by subsurface thermal measurement. However, the complexity of these phenomena can make measurements tricky because it is both far from the surface and relatively simple. It has been shown that a prediction of squeal is achievable by processing both the temperature level and the temperature difference between two measurements located in front of and behind the direction of sliding, provided that the loading history is considered, in particular the high-temperature events which transform the pad material. Discrete surface analysis has shown that the temperature level affects the tribolayer and the difference between the measurements affects the macroscopic load-bearing variation. This type of indicator seems to be able to reflect the intricate mechanisms governing brake system behavior.

Different improvements can be suggested as future work. In-depth processing of squealing events and frequency classification could explain at least in part the overlaps of the fields observed in Figure 13. The number of sensors could be increased, particularly thermal, to access finer location information. Data processing is an important area of improvement, whether by adding information from sensors other than thermal or by using more systematic processing such as machine learning methods, or even by extended processing of measurements to go back to surface information (inverse models). Finally, while, in this study, the focus has been placed on the pin, extensive observation of the disc's surface would be necessary to have a more complete comprehension of the contact behavior.

Various applications could be found for this work. The insights derived from this study could inform the development of strategies to mitigate squeal occurrences by addressing factors such as temperature mitigation. Conditions of brake use that are conducive to squeal could be defined and avoided, for example, with the design of the brake or with the combination of other brakes, such as regenerative brakes.

Author Contributions: M.T. conducted the experiments, analyzed the results, and wrote the manuscript. J.-F.B. conceived the experiment and participated in the writing process. F.B. developed the test bench control and data acquisition system. P.D. contributed to the concept and the methodology of the research, led the project, and participated in the writing process. M.B. assisted with surface analysis. M.S. assisted with data analysis. N.H. contributed to the concept and the methodology of the research and led the project. All authors have read and agreed to the published version of the manuscript.

Funding: This work was carried out as part of the Franco-German PI-CUBE project financed by the BMBF-DFG and MESRI-ANR institutions. It has also been supported by the RITMEA project, which is co-financed by the European Union with the European Regional Development Fund, the French state and the Hauts de France Region Council.

Data Availability Statement: The data generated during and/or analyzed during the study are available from the corresponding author upon reasonable request.

Conflicts of Interest: The authors declare no competing interests.

References

1. Mackin, T.J.; Noe, S.C.; Ball, K.J.; Bedell, B.C.; Bim-Merle, D.P.; Bingaman, M.C.; Bomleny, D.M.; Chemlir, G.J.; Clayton, D.B.; Evans, H.A.; et al. Thermal cracking in disc brakes. *Eng. Fail. Anal.* **2002**, *9*, 63–76. [CrossRef]
2. Österle, W.; Dörfel, I.; Prietzel, C.; Rooch, H.; Cristol-Bulthé, A.L.; Degallaix, G.; Desplanques, Y. A comprehensive microscopic study of third body formation at the interface between a brake pad and brake disc during the final stage of a pin-on-disc test. *Wear* **2009**, *267*, 781–788. [CrossRef]
3. Massi, F.; Berthier, Y.; Baillet, L. Contact surface topography and system dynamics of brake squeal. *Wear* **2008**, *265*, 1784–1792. [CrossRef]
4. Kinkaid, N.M.; O'Reilly, O.M.; Papadopoulos, P. Automotive disc brake squeal. *J. Sound Vib.* **2003**, *267*, 105–166. [CrossRef]
5. Akay, A. Acoustics of friction. *J. Acoust. Soc. Am.* **2002**, *111*, 1525–1548. [CrossRef]
6. Hoffmann, N.; Fischer, M.; Allgaier, R.; Gaul, L. A minimal model for studying properties of the mode-coupling type instability in friction induced oscillations. *Mech. Res. Commun.* **2002**, *29*, 197–205. [CrossRef]
7. Eriksson, M.; Jacobson, S. Friction behaviour and squeal generation of disc brakes at low speeds. *Proc. Inst. Mech. Eng. Part D J. Automob. Eng.* **2001**, *215*, 1245–1256. [CrossRef]
8. James, S.; Ouyang, H.; Brookfield, D.J.; Mottershead, J.E. Disc Brake Squeal—An Experimental Approach. *Mater. Sci. Forum* **2003**, *440–441*, 237–244. [CrossRef]
9. Duboc, M.; Magnier, V.; Brunel, J.F.; Dufrénoy, P. Experimental set-up and the associated model for squeal analysis. *Mech. Ind.* **2020**, *21*, 204. [CrossRef]
10. Bergman, F.; Eriksson, M.; Jacobson, S. The effect of reduced contact area on the occurrence of disc brake squeals for an automotive brake pad. *Proc. Inst. Mech. Eng. Part D J. Automob. Eng.* **2000**, *214*, 561–568. [CrossRef]
11. Xiang, Z.Y.; Mo, J.L.; Ouyang, H.; Massi, F.; Tang, B.; Zhou, Z.R. Contact behaviour and vibrational response of a high-speed train brake friction block. *Tribol. Int.* **2020**, *152*, 106540. [CrossRef]
12. Hetzler, H.; Willner, K. On the influence of contact tribology on brake squeal. *Tribol. Int.* **2012**, *46*, 237–246. [CrossRef]
13. Bergman, F.; Eriksson, M.; Jacobson, S. Influence of disc topography on generation of brake squeal. *Wear* **1999**, *225–229*, 621–628. [CrossRef]
14. Sinou, J.J.; Jézéquel, L. Mode coupling instability in friction-induced vibrations and its dependency on system parameters including damping. *Eur. J. Mech. A/Solids* **2007**, *26*, 106–122. [CrossRef]
15. AbuBakar, A.R.; Ouyang, H. Wear prediction of friction material and brake squeal using the finite element method. *Wear* **2008**, *264*, 1069–1076. [CrossRef]
16. Belhocine, A.; Ghazaly, N.M. Effects of material properties on generation of brake squeal noise using finite element method. *Lat. Am. J. Solids Struct.* **2015**, *12*, 1432–1447. [CrossRef]
17. Hentati, N.; Kchaou, M.; Cristol, A.L.; Najjar, D.; Elleuch, R. Impact of post-curing duration on mechanical, thermal and tribological behavior of an organic friction material. *Mater. Des.* **2014**, *63*, 699–709. [CrossRef]
18. Godet, M. The third-body approach: A mechanical view of wear. *Wear* **1984**, *100*, 437–452. [CrossRef]
19. Dowson, D.; Taylor, C.; Childs, T.; Dalmaz, G.; Berthier, Y.; Flamand, L.; Georges, J.; Lubrecht, A. *The Third Body Concept: Interpretation of Tribological Phenomena*; Elsevier: Amsterdam, The Netherlands, 1996; Volume 31.
20. Denape, J. Third body concept and wear particle behavior in dry friction sliding conditions. *Key Eng. Mater.* **2015**, *640*, 1–12. [CrossRef]
21. Davin, E.; Cristol, A.L.; Brunel, J.F.; Wear, Y.D.; Brunel, J.F.; Desplanques, Y. Wear mechanisms alteration at braking interface through atmosphere modification. *Wear* **2019**, *426*, 1094–1101. [CrossRef]
22. Kchaou, M.; Lazim, A.R.M.; Hamid, M.K.A.; Bakar, A.R.A. Experimental studies of friction-induced brake squeal: Influence of environmental sand particles in the interface brake pad-disc. *Tribol. Int.* **2017**, *110*, 307–317. [CrossRef]
23. Tison, T.; Heussaff, A.; Massa, F.; Turpin, I.; Nunes, R.F. Improvement in the predictivity of squeal simulations: Uncertainty and robustness. *J. Sound Vib.* **2014**, *333*, 3394–3412. [CrossRef]
24. Kasem, H.; Bonnamy, S.; Berthier, Y.; Jacquemard, P. Characterization of surface grooves and scratches induced by friction of C/C composites at low and high temperatures. *Tribol. Int.* **2010**, *43*, 1951–1959. [CrossRef]
25. Lee, S.; Jang, H. Effect of plateau distribution on friction instability of brake friction materials. *Wear* **2018**, *400–401*, 1–9. [CrossRef]
26. Lai, V.V.; Paszkiewicz, I.; Brunel, J.F.; Dufrénoy, P. Multi-scale contact localization and dynamic instability related to brake squeal. *Lubricants* **2020**, *8*, 43. [CrossRef]
27. Deltombe, R.; Kubiak, K.J.; Bigerelle, M. How to select the most relevant 3D roughness parameters of a surface. *Scanning* **2014**, *36*, 150–160. [CrossRef]

lubricants

MDPI

Article

International Perspectives on Skid Resistance Requirements for Pavement Markings: A Comprehensive Synthesis and Analysis

Jieyi Bao [1], Hua Zhao [2], Yi Jiang [3] and Shuo Li [4,*]

[1] Stock Development Department of Construction Management, Florida Gulf Coast University, Fort Myers, FL 33965, USA; jbao@fgcu.edu
[2] School of Infrastructure Engineering, Nanchang University, Nanchang 330031, China; zhaohua@ncu.edu.cn
[3] School of Construction Management Technology, Purdue University, West Lafayette, IN 47907, USA; jiang2@purdue.edu
[4] Research Division, Indiana Department of Transportation, Indianapolis, IN 46204, USA
* Correspondence: sli@indot.in.gov

Abstract: Pavement (or road) markings play an important role in road safety, influencing the dynamics of road users through their skid resistance properties. This study provides a comprehensive synthesis and analysis of international perspectives on the skid resistance of pavement markings and their requirements. It examines marking skid test results across various regions, including North America, Europe, and other parts of the world, and emphasizes the impact of different materials and test environments on skid resistance. The study also reviews current skid resistance standards and guidelines, from North American state-level standards to European and global specifications. Furthermore, it discusses the safety implications of these standards for diverse road users, especially motorcyclists, cyclists, and pedestrians. In conclusion, this paper highlights the importance of further innovation and consistency in skid resistance testing and standards to improve road safety.

Keywords: pavement/road marking; skid resistance/friction; skid/friction number; British pendulum number; macrotexture; mean profile depth

1. Introduction

Pavement or road markings are essential components of traffic control systems, providing guidance and safety for road users [1]. Traditionally, the performance of these markings, often enhanced with glass beads for improved retro-reflectivity, has been evaluated primarily based on visibility factors like retro-reflectivity and color. However, marking materials often fill small voids in the road surface, and the smooth, round shape of these glass beads alters the surface texture of the pavement. This leads to a reduction in skid resistance, particularly in wet conditions. On average, the skid resistance of pavement markings is 15–20% lower than that of the surrounding pavement surface [2].

While Europe [3] has long considered skid resistance a key performance requirement for pavement markings, this issue has not received the same level of attention in other regions. In the United States, for instance, the Manual on Uniform Traffic Control Devices (MUTCD) [1] acknowledges the need to minimize tripping hazards and the loss of traction but lacks clear, practical implementation guidelines. As a result, current evaluations of pavement markings tend to prioritize visibility over skid resistance, with limited research on the skid resistance of pavement markings and its safety implications.

The skid resistance of pavement markings depends on various factors, including the type of materials (such as binders and glass beads), the design of the markings (e.g., thickness and profile), the pavement surface texture, and environmental conditions like moisture and temperature. Advances in high-performance glass beads and retro-reflective technologies have enhanced visibility, particularly in low-light conditions and adverse weather [4,5]. However, the application of more drop-on glass beads increases retro-reflectivity but may reduce skid resistance [6]. While engineered anti-skid additives have been developed to improve skid resistance [7–11], achieving both high retro-reflectivity and high skid resistance remains challenging.

Concerns about the skid resistance of pavement markings arise commonly in two situations: poor friction and differential friction [12–14]. Both can lead to the loss of control, especially during turning maneuvers. Such risks are increased in high-traffic areas like intersections, where large markings such as crosswalks, letters, and arrows are common. Moreover, the advent of autonomous vehicles has led to recommendations for wider longitudinal markings [1,15]. This paper presents a comprehensive synthesis of the skid resistance of pavement markings, identifies research gaps, and discusses practical limitations, aiming to inform future studies and refine methodologies to improve roadway safety concerning the skid resistance of pavement markings. Note that, for the purpose of this paper, terms such as "road markings" and "pavement markings", as well as "friction" and "skid resistance", are used interchangeably to refer to the same concepts.

2. Pavement Markings: Types and Materials

2.1. Pavement Markings

Pavement markings consist of a binder (resin), pigments, fillers, and sometimes a dispersion medium [16]. The binder ensures adhesion and cohesion, while pigments provide color, with titanium dioxide (TiO_2) and lead chromate (PbCrO) being common for white and yellow markings [17]. Fillers, typically composed of a mixture of different materials, including calcium carbonate, sand, calcined clays, silicates, mica, barium sulfate, and other materials, enhance markings by improving moisture resistance, abrasion durability, temperature stability, visibility, and visual appearance. Some materials, like solvent-borne paints, include a dispersion medium to maintain stability, distribute components, and aid film formation.

2.1.1. Paints

Solvent-borne paints are single-component coatings consisting of binder resin, pigments, fillers, and solvents [18,19]. Their classification depends on the binder type, such as epoxy, alkyd, or acrylic. These paints contain about 30% solvent, which evaporates during application, forming a solid film. Though durable and weather-resistant, solvent-borne paints have a service life of 6–12 months and are increasingly replaced by waterborne alternatives due to environmental concerns and higher costs. They are used on clean, dry surfaces with a wet film thickness of 15 mils (1 mil = 0.0254 mm).

Waterborne paints, comprising binders, pigments, fillers, and water-based solvents, account for nearly 90% of pavement markings in the U.S. [17]. These paints dry faster, are less influenced by humidity, and offer strong adhesion to asphalt and concrete surfaces [20,21]. With a typical wet film thickness of 15–25 mils, waterborne paints last 9–36 months. While more durable than solvent-borne paints, they are best suited for low-traffic or temporary markings due to their relatively lower longevity.

2.1.2. Multi-Component Paint

Epoxy is a two-component marking material consisting of two parts: component A, containing resin, pigments, extenders, fillers, and solvents, and component B, the catalyst that accelerates the curing process [16]. Proper surface preparation is critical for achieving optimal bonding when applying epoxy. The mixing ratio of component A to component B is typically 1:1 or 2:1 by volume. Once mixed, epoxy undergoes an exothermic reaction, i.e., releasing heat (or energy). Epoxy markings can be applied to either a concrete or asphalt surface. As a durable marking, the lifespan of epoxy markings exceeds four years on roads with low to medium traffic volumes [20,21]. The applied film thickness of epoxy markings typically ranges from 15 mils to 25 mils. Epoxy paints can be used for both longitudinal and transverse markings.

Polyurea is also a two-component system. Component A is a resin mixture containing pigments and fillers, while component B serves as a catalyst [16], typically mixed at a 2:1 ratio by volume. Unlike epoxy, polyurea undergoes an endothermic reaction, requiring energy input, and solidifies rapidly within seconds. Polyurea is a durable marking, with a service life of up to five years. Due to its benefits such as its fast-drying, abrasion-resistant, and high-performance properties, polyurea is a popular choice for both high-traffic and industrial environments. The recommended wet film thickness for polyurea pavement markings is 15 mils to 25 mils [21].

Methyl methacrylate (MMA) is a solvent-free, two-component system [16]. Component A contains a methyl methacrylate monomer, pigments, and fillers, and component B consists of a liquid or powder catalyst. Typical mixing ratios are 4:1 or 1:1 by volume, and 98:2 by weight, depending on the intended function of the marking. Aggregates are incorporated to enhance skid resistance. When MMA is applied to pavement surfaces, it forms a strong bond through an exothermic reaction. MMA typically lasts more over 3 years [22] and is widely used in bicycle and dedicated bus lanes. The application thickness of MMA varies according to the vendor requirements, formula, and intended use, commonly ranging from 10 to 250 mils [23].

2.1.3. Thermoplastics

Thermoplastic (TP) markings consist of a blend of resin, fillers, and pigments, with hydrocarbon and alkyd being the most used resins [16]. TP materials are solvent-free and available in two forms: dried palletized and preformed-shaped. The former requires heated extruding, where the material is melted and applied to the pavement surface, while the latter requires heating the material with a torch. TP markings are commonly melted at temperatures of 180 °C to 220 °C [17]. TP markings exhibit resistance to snowplow damage, and can be applied over existing markings, eliminating the need for prior removal. However, their application is generally limited to asphalt surfaces. TP markings can last 3 to 6 years, and their application thickness ranges from 90 mils to 125 mils.

2.1.4. Tapes

Pavement marking tapes include two types: permanent and temporary tapes. Permanent preformed tapes are typically made from plastic binder materials, with urethane and pliant polymer being the two most used binders for permanent tapes [21]. Temporary marking tapes are composed of synthetic polymer [24]. Marking tapes are typically supplied in continuous rolls and are manufactured with embedded glass beads or other particles to enhance reflectivity and retro-reflectivity. When installed correctly, permanent preformed tapes can have a service life of 4 to 8 years.

2.2. *Reflective Materials and Particles*

Glass beads are applied directly onto freshly applied pavement markings or, in certain cases, partially mixed into markings prior to application (pre-mixing paint) [16]. The AASHTO M247-13 specification [25] divides glass beads into six types based on gradations (see Table 1). Glass beads are available in two types: coated and uncoated. Coated (or treated) glass beads possess a surface coating that facilitates their embedding into the paint, whereas uncoated beads remain on the surface. The index of refraction (IOR) of glass beads ranges from 1.50 to 1.55. Ceramic particles, angular particles, and aggregates have also been developed to enhance the skid resistance of pavement markings [7]. Due to the absence of a standardized specification, these materials may exhibit variations in size and shape.

Table 1. Gradations of glass beads.

Sieve Size (mm)	Mass % Passing					
	Type 0	Type 1	Type 2	Type 3	Type 4	Type 5
2.35						100
2.00					100	95–100
1.70				100	95–100	80–95
1.40				95–100	80–95	10–40
1.18		100	100	80–95	10–40	0–5
1.00				10–40	0–5	0–2
0.850		95–100	90–100	0–5	0–2	
0.600	100	75–95	50–75	0–2		
0.425	90–100		15–45			
0.300	50–75	15–35	0–15			
0.180	0–5		0–5			
0.150		0–5				

As the end of this section, Table 2 summarizes the application aspects of different pavement markings reviewed above.

Table 2. Application aspects of different pavement markings from reviewed references.

Marking Type	Thickness (mils) [1]	Service Life	Application [2]	Advantages and Disadvantages
Solvent-borne	15	6–12 months	• Longitudinal and transverse markings; • Low traffic volume (e.g., AADT [3] < 3000).	Advantages: • Low cost; • Fast-drying; • Weathering resistance; • Suitable for asphalt and concrete. Disadvantages: • Environmental contamination risk; • Health hazards for workers; • Flammable; • Low durability.
Waterborne	15–25	9–36 months	• Longitudinal and transverse markings; • Low traffic volume (e.g., AADT < 3000).	Advantages: • Low cost; • Fast-drying; • Less affected by humidity; • Suitable for asphalt and concrete; • Environmentally friendly; • Stable properties; • Easy to apply. Disadvantages: • Low durability.
Epoxy	15–25	4 years	• Longitudinal and transverse markings; • Medium traffic volume (e.g., AADT ≥ 3000).	Advantages: • Suitable for asphalt and concrete; • Durable; • Environmentally friendly. Disadvantages: • Prone to yellowing or fading under UV exposure; • Incompatible with existing markings of other materials; • Slow-drying.

Table 2. *Cont.*

Marking Type	Thickness (mils) [1]	Service Life	Application [2]	Advantages and Disadvantages
Polyurea	15–25	5 years	• Longitudinal markings; • Medium traffic volume (e.g., AADT ≥ 3000).	Advantages: • Fast drying; • Resistant to abrasion, chemicals, and UV radiation; • Durable and moisture-resistant; • Applicable in freezing temperatures; • Suitable for asphalt and concrete; • Environmentally friendly. Disadvantages: • Requires specialized equipment for application.
MMA	10–250	3 years	• Bicycle and bus lanes; • Longitudinal and transverse markings; • Symbols; • Medium traffic volume (e.g., AADT ≥ 3000).	Advantages: • Durable; • Environmentally friendly; • Suitable for asphalt and concrete. Disadvantages: • Requires dry conditions for application; • Produces objectionable odor.
Thermoplastics	90–125	3–6 years	• Longitudinal and transverse markings; • Symbols; • Medium traffic volume (e.g., AADT ≥ 3000).	Advantages: • Durable; • Applicable on existing markings; • Resistant to snowplow damage. Disadvantages: • Limited to asphalt surfaces; • Requires heat or specialized devices.
Temporary Tapes	Multiple thicknesses	1–6 months	• Work zones or areas need temporary traffic control.	Advantages: • Easy to remove; • Simple application. Disadvantages: • Short service life; • High costs compared to paint.
Permanent Preformed Tapes	Multiple thicknesses	4–8 years	• Longitudinal and transverse markings; • Symbols; • High traffic volume (e.g., AADT ≥ 18,000).	Advantages: • Durable; • Suitable for asphalt and concrete surfaces; • Easy to apply; • Roads can reopen immediately. Disadvantages: • High costs.

[1] 1 mil = 0.0254 mm; [2] See reference [26]; and [3] AADT is annual average daily traffic.

3. Skid Resistance Test Results of Pavement Markings

3.1. North America

In 1975, Richard [27] conducted a two-phase study on the skid resistance of pavement markings in the U.S. Phase I tested 14 marking types in the lab using the British pendulum tester (BPT), with results ranging from a British pendulum number (BPN) of 45 for fast-dry white paint (no beads) to 14 for smooth white cold plastic. Phase II compared three marking materials on a freeway, measuring the field friction number (FN40 at 40 mph) against lab BPN data. Fast-dry beaded white paint had an FN40 of 37 and a BPN of 31, while extruded yellow hot plastic showed an FN40 of 23 and a BPN of 35. The study concluded that pavement markings generally have lower friction than road surfaces and recommended using abrasive additives to improve skid resistance.

In 1980, Anderson and Henry [12] tested 39 formulations from 11 types of marking materials, including paints, thermoplastics, cold-applied plastics, and temporary tapes, under lab and field conditions. They measured wet friction using the skid number (SN), BPN, microtexture, macrotexture, and static friction coefficients. The results showed a significant variation in wet friction among materials. Beaded materials had a consistent BPN of 50 ± 5, with little change after polishing, as beads mainly influenced friction. Unbeaded materials, like chlorinated rubber paints and thermoplastics, had lower and more variable friction. Thin materials, such as paints, lost skid resistance over time, while thick materials, like thermoplastics, did not recover skid resistance after prolonged exposure. The study highlighted safety risks from uneven skid resistance between marked and unmarked areas, especially over large zones.

In 1995, Bagot [28] tested five materials: two epoxies, two waterborne paints, and one MMA resin, at three airports to identify durable, visible options that reduce maintenance costs and meet environmental standards. Using the K.J. Law Runway Friction Tester (RFT), he found friction values ranging from 0.32 to 0.90 for epoxy, 0.44 to 0.99 for waterborne paint, and 0.31 to 0.68 for MMA, depending on the use of silica granules. The study concluded that silica additives improved skid resistance in epoxies. In 1996, Bagot [29] furthered friction tests to assess the effects of retro-reflective glass beads and silica. Adding 1.5 and 1.9 IOR beads increased initial friction from 0.54 to 0.64 and 0.63, respectively, while adding silica raised it to 0.84. Silica enhanced friction the most, but its effectiveness declined when combined with beads due to size differences.

Rodin et al. [22] investigated the safety of different marking materials for motorcyclists and bikers in 2018. Three types of markings, including waterborne paint, TPs, and preformed tape, were assessed for friction in dry, wet, and iced conditions. In the laboratory, beaded paint exhibited the highest BPN at 100, and beaded TPs exhibited the lowest BPN at 62 in dry conditions, and, in wet conditions, preformed tape exhibited the highest BPN at 64, and beaded TPs exhibited the lowest BPN at 40. The results also showed that the wet BPNs measured with a pedestrian slip rubber and a tire slip rubber are very close.

Fanijo et al. [30] evaluated waterborne paint, MMA, and TP markings for bicycle lanes under simulated wear conditions in 2023. Friction measurements included the mean texture depth (MTD) and the international friction index (IFI). The results showed that TP markings had a higher MTD (1.20 mm) than beaded paint (0.90 mm), unbeaded paint (0.59 mm), and MMA (0.62 mm). Before polishing, MMA had the highest IFI at 0.40, while unbeaded waterborne paint had the lowest IFI at 0.18. The test results also showed that friction decreased for MMA and slightly increased for the waterborne paint, after polishing.

The latest research by Bao et al. (2024) examined the skid resistance of six beaded markings [31]. Laboratory tests revealed the following mean profile depth (MPD) and BPN values: beaded waterborne (MPD 0.31–1.26 mm, BPN 40.0–62.5), preformed tapes (MPD 1.08 mm, BPN 40.8), epoxy (MPD 0.42–0.75 mm, BPN 33.8–35.5), polyurea (MPD 0.37–1.55 mm, BPN 34.0–42.5), MMA (MPD 0.34 mm, BPN 47.5), and thermoplastics (MPD 0.25–0.26 mm, BPN 30.5–32.5).

3.2. Europe

In the United Kingdom (UK), Reid et al. (1962) [32] studied white-line road markings and light-colored surfaces. Laboratory tests showed that TP markings with glass beads had a skid resistance of 37–58 BPN without gritty aggregate and 48–90 BPN with it. A survey of 100 road sites found that all TP markings had a skid resistance of at least 55 BPN, with most exceeding 65 BPN, while three paints fell below 55 BPN. TP markings were the most durable and skid-resistant. Purohit et al. (2020) [33] assessed preformed 3D TP markings on asphalt and concrete. On Day 1, MTD values were 4.0 mm on asphalt and 2.8 mm on concrete, with slight reductions after a year. The initial BPN values on asphalt were 84.2 (dry) and 69.0 (wet), dropping slightly to 81.3 (dry) and 68.0 (wet) after a year. On concrete, the initial BPN values were 90.4 (dry) and 73.0 (wet), decreasing to 82.5 (dry) and 69.0 (wet) after a year. The markings maintained a stable texture and high friction over 12 months.

In Italy, Pasetto and Manganaro (2006) [9] studied the impact of surface texture saturation on skid resistance for solvent-based paints with glass beads and anti-skid granules (80:20 weight ratio) applied at a depth of 300 μm. The markings had skid resistance tester values (SRT) of 41–55 and an MTD of 0.51–0.91 mm. Skid resistance decreased by 15–20%, and macrotexture by 10%, with greater effects on surfaces with an initially high texture. Asdrubali et al. [34] evaluated road markings at 28 sites using various materials, including paint, TP, two-component, and preformed markings. Tests conducted in 2010 and 2012 showed SRT values ranging from 31–63 in 2010 and 32–68 in 2012.

Lundkvist and Isacsson (2007) carried out wet nighttime measurements in three test-fields, two in Sweden and one in Demark [35]. Skid resistance and texture were measured on 130 flat TP road markings. The results showed that the MPD ranged between 0.15 mm and 2.35 mm, and the SRT ranged from 0.63 to 0.92. The findings indicated that a dry texture could predict wet skid resistance using models validated through regression analysis.

Coves-Campos et al. (2018) [7] tested 18 road marking types on a rural highway in Spain using styrene acrylic paint, four types of glass beads (125–1180 μm), and two anti-skid aggregates (marble sands and calcium-sodium granules). Initial SRT values ranged from 49.0 to 55.1 for markings with glass beads alone, 58.0–63.0 with 80% glass beads and 20% marble sands, 52.0–55.2 with 80% glass beads and 20% granules, and 53.0–58.5 for double-layer systems using beads or granules. After 18 months, SRT values declined to 41.2–45.1, 44.0–49.0, 43.0–46.5, and 43.5–48.5, respectively. Glass beads offered better visibility but lower durability, anti-skid aggregates improved skid resistance at the cost of visibility, and double-layer systems demonstrated the best overall durability.

In Austria, Burghardt et al. (2023) [8] studied how glass beads and anti-skid particles affect the skid resistance of pavement markings. They tested five setups: bare asphalt, paint only, paint with glass beads, paint with glass beads and 10% corundum, and paint with anti-skid particles. Waterborne paints, 400 μm thick, were applied as transverse stripes on a road with low-speed traffic (400 trucks and 200 cars daily). The initial and 10-month SRT values were as follows: asphalt (49.0 to 47.0), paint only (35.0 to 36.0), paint with glass beads (44.0 to 41.0), paint with glass beads and corundum (50.0 to 42.0), and paint with anti-skid particles—corundum (48.0 to 45.0), glass granulates (65.0 to 44.0), cristobalite (53.0 to 46.0), and bauxite (61.0 to 49.0). The study found that adding anti-skid particles is critical to reducing slipperiness in thin markings.

3.3. Other Regions

Thew and Dabic (2000) evaluated the skid resistance of three paints and adjacent road surfaces in Australia [36]. The results exhibited that waterborne paints with Type C beads demonstrated a skid resistance that was 9–13 BPN lower than that of the adjacent road surface, while alkyd markings exhibited a skid resistance that was 19–22 BPN lower than that of the surrounding road surfaces. Drop-on glass beads increased the skid resistance, typically from 25–30 BPN to 35–38 BPN.

In South Africa, Naidoo and Steyn (2018) [37] tested various road marking materials, including white and yellow waterborne, 1.2 mm TP, 1 mm cold plastic, and 3 mm screed materials. Skid resistance was evaluated using test markings on plates in the lab. White waterborne, 1 mm cold plastic, and both white and yellow 1.2 mm TPs had an SRT value greater than 50. Other materials had SRT values between 35 and 45. The study suggested that road marking applicators should increase the amount of anti-skid aggregate to improve skid resistance.

Siyahi et al. (2015) [38] studied the effect of additives (ground waste glass, silica, and Lika powders) on the properties of a two-component acrylic paint used in Iran. They applied 800-micron thick paint samples on asphalt slabs. Skid resistance for the paint without additives was 33 BPN, and it increased by 46%, 33%, and 25% when 5% Lika, silica, and waste glass powders were added, respectively. Hadizadeh et al. (2020) [39] evaluated MMA-based cold plastic traffic paints under simulated conditions, with samples applied to degreased steel panels. Skid resistance ranged from 48 to 74 BPN using two different aggregates of silica.

In Mainland China, Wang et al. (2023) [40] evaluated tire-road wear using a two-wheel accelerated wear test with three specimens: 13 mm hot mix asphalt, unbeaded solvent-based paint, and beaded solvent-based paint. Initial BPN values were 54 for asphalt, 35 for unbeaded paint, and 70 for beaded paint. After 150 min, the BPN values changed to 61, 53, and 54, respectively. Chen et al. (2020) [10] tested traditional and modified hot-melt paints with glass beads, using a four-wheel accelerated polishing machine. The initial BPN was 65 for traditional paint and 68 for modified paint. After 10,000 cycles, the BPN decreased to 51 and 54. Yang (2020) [11] also reported a high-performance highway marking with a BPN of 65.

In Taiwan, Chiu et al. (2017) [41] studied heat-treated polyester markings on two highways. White beaded longitudinal markings, 15 cm wide, were applied to new asphalt pavements. The initial BPN values ranged from 45 to 68 at 18 test points, but half of the points showed a drop below 50 BPN after one year. Su et al. (2021) [42,43] assessed the skid resistance of marking materials, focusing on those with 65 BPN or higher. At real intersections, 65 BPN markings, aggregate markings, and cold plastic markings were tested. The initial BPN for 65 BPN markings was 60, dropping to 43 after 400 days and increasing to 58 after 575 days. Aggregate markings started at 52 BPN after 65 days and rose to 79 after 340 days. Cold plastic markings began with 95 BPN and dropped to 55 after one year. Lab tests showed that 65 BPN markings started at 71 BPN, decreasing to 59 after 150,000 polishing cycles.

4. Skid Resistance Requirements for Pavement Markings

4.1. North America

The MUTCD recommends selecting pavement marking materials that reduce the risk of tripping or losing traction for all road users, including pedestrians, cyclists, and motorcyclists. Florida Department of Transportation (DOT) [44] mandates a minimum skid resistance of 35 BPN for pavement markings, and 55 BPN for bicycle markings and crosswalks. Texas DOT [45] requires high-build paints to have an initial skid resistance of at least 45 BPN. Georgia DOT [46] sets a minimum of 35 BPN for preformed plastic markings. Illinois DOT [47] requires at least 45 BPN for Types B and D materials (both with patterned surfaces, intermixed glass beads with IOR $\geq$ 1.50, and top-coated ceramic particles with IOR between 1.80 and 1.70) and 55 BPN for Type C. The surface of blackout pavement marking tape must also have a minimum of 45 BPN. Other state DOTs [48–51] require at least 45 BPN for preformed tapes and TP markings. Municipal agencies [52–54] require preformed TP markings to have an initial BPN of 55 or 60 and maintain at least 45 BPN.

4.2. Europe

The EN standard, EN 1436 [3], which addresses marking performance, serves as the basis for establishing skid resistance requirements for road markings in EU member states. The skid resistance of pavement markings is measured as the skid resistance tester value (SRT) that is divided into six classes of S0 to S5. Class S0 denotes situations where no SRT is requested or when the SRT cannot be measured, while Classes S1 to S5 denote markings with SRT $\geq$ 45, 50, 55, 60, and 65, respectively. The permitted skid resistance classes range from S1 to S5.

In the UK [55], the minimum skid resistance is 55 for critical areas and 45 for non-critical areas. In France [56], the Decree of 10 May 2000 sets a minimum Class S1 (SRT $\geq$ 45) for all markings, with a recommended Class S3 (SRT $\geq$ 55) for areas requiring higher grip, such as pedestrian crossings. Nordic countries [57] require friction values of at least 0.52 for type I and II markings, temporary markings, and durable markings; 0.65 for hand-applied retro-reflective markings; and 0.71 for hand-applied non-retro-reflective and anti-skid markings. Poland [58] requires SRT values of 50 for motorways and express roads and 45 for other roads.

4.3. Other Regions

In New Zealand, NZTA M7 [59] requires a minimum BPN of 45 for markings with a thickness <0.9 mm and 50 for markings with a thickness $\geq$0.9 mm. NZTA M20 [60] requires a skid resistance of 50–65 BPN for long-lasting markings one hour after application and beyond. In Australia, the skid resistance is specified for TP markings (at a minimum of 45 BPN) [61] and high-performance markings [62], which are classified into three categories: no requirement (SK0), 45–60 BPN (SK1), and over 60 BPN (SK2).

In Mainland China, the initial skid resistance for all markings must be at least 45 BPN [63], with an upcoming revision [64] to set 45 BPN for conventional markings and 55 BPN for anti-skid markings. Taiwan's specifications require a minimum initial BPN of 50 for TP markings [65] and classify markings into six classes [66], requiring a minimum SRT of 45. Tung (2020) [67] reviewed the skid resistance specifications and found that many countries, including Singapore, Vietnam, Malaysia, Indonesia, and India, set the minimum BPN threshold at 45.

5. Discussion and Analysis

5.1. Safety Demand

Pavement markings are installed on road surfaces, so skid resistance requirements for markings often align with those for the road. Kummer and Meyer (1967) [68] recommended a minimum pavement friction of 37 SN40R (test speed: 40 mph, rib tire). Zhao et al. (2020) [69] linked the AASHTO [70] deceleration threshold (3.4 m/s^2) to a minimum friction coefficient of 0.35. Shuo et al. (2021) [71] found that 37 SN40R equals 20 SN40S (smooth tire) and that setting a minimum above 20–23 SN40S could raise maintenance costs significantly.

Friction demand and handling differ significantly for four-wheel and two-wheel vehicles. Four-wheel vehicles, with larger tire footprints, are more stable, whereas motorcycles are more affected by pavement friction. Research [72,73] has mainly focused on motorcycle sliding friction for accident reconstruction rather than the friction at the moment of a crash. Bicycles, with unique dynamics, generally require less friction than motorcycles but need enough for safe stopping and control. In the U.S., AASHTO [74] specifies friction coefficients of 0.32 (dry) and 0.16 (wet) for bicycle lanes. South Australia [75] mandates a grip number (GN, where GN = 0.01 $\times$ BPN) of 0.40 for bikeways, Korea [76] requires a 40 BPN, and Andalusia [77] sets a 0.25 friction coefficient for safe stopping distances on paved roads.

Pedestrians, though not as speed-dependent as motorcycles and bicycles, still require sufficient friction to avoid slips and falls, especially in busy areas or adverse conditions. ASTM D2047 [78] recommends a static friction coefficient of 0.50 for floor surfaces. However, slip risks on roads, floors, and work surfaces vary due to differences in the environment, users, and consequences. In Japan, Yamada et al. [79] suggested a slip friction coefficient of 0.34 (31 BPN) for wet conditions. Tanaka and Uchida [80] identified surfaces with 40 BPN or lower as slippery. Miyata et al. [81] found no significant benefit above 40 BPN, proposing it as the standard, and that a BPN of 30 is a critical safety threshold, with values below 30 indicating risk and 20 considered hazardous.

Table 3 summarizes pavement marking performance requirements from various global roadway agencies, as discussed in Section 4. Most standards or specifications prioritize retro-reflectivity and color for visibility, while skid resistance requirements are absent in the U.S., Japan, and Korea standards. Note that the two main BPT test methods, EN 13036 [82] and ASTM E303 [83], show slight differences [41]. Seemingly, a minimum skid resistance of 40 BPN may be acceptable for low-speed, low-traffic areas like residential streets or rural roads. However, for high-speed, high-traffic, or critical areas such as crosswalks, intersections, or regions prone to wet or icy conditions, a minimum of 45 BPN or higher is recommended to ensure sufficient skid resistance.

Table 3. Minimum requirements for pavement marking performance from reviewed references.

Source	Reflection	Retro-Reflection	Color	Durability	Skid Resistance [1]
US: MUTCD [1]	No	Yes	No	No	No
INDOT [84]	No	Yes	Yes	Yes	No
Canada: Ontario MTO [85]	No	Yes	Yes	Yes	No
UK: BSI [86]	Yes	Yes	Yes	Yes	45 (55 [2])
EU: EN 1436 [3]	Yes	Yes	Yes	Yes	45
New Zealand: <0.9 mm thick [59]	Yes	Yes	Yes	Yes	45
≥0.9 mm thick [59]	Yes	Yes	Yes	Yes	50
Long-life [60]	Yes	Yes	Yes	Yes	≥50 and ≤65
Australia: Paints [87,88]	Yes	Yes	Yes	Yes	No
Thermoplastic [61]	Yes	Yes	Yes	Yes	40 (initial)
High-Performance [62]	Yes	Yes	Yes	Yes	No/45–60/>60
Mainland China: GB/T 16311 [63]	No	Yes	Yes	No	45
GB/T 16311 pending [64]	No	Yes	Yes	No	45/55
Taiwan: T.E. Specs. [65]	Yes	Yes	Yes	Yes	≥50 (TP)
CNS 15834 [66]	Yes	Yes	No	No	45~65
Japan: JCASM [89]	Yes	Yes	Yes	No	No
Korea: KS M6080 [90]	Yes	Yes	Yes	No	No

[1] Skid resistance is in BPN or SRT; and [2] Required for critical locations.

5.2. Engineering and Technical Feasibility

Establishing minimum friction requirements for pavement markings involves balancing material science, engineering, and economics. The choice of materials, such as TPs, epoxy resins, high-friction aggregates, and additives, is key to ensuring durability and adequate texture for marking skid resistance. Application methods like spraying, rolling, or using preformed tape influence the texture and performance, each with its pros and cons. Moreover, regular maintenance helps maintain skid resistance over time. While the initial costs may vary significantly by material (see Table 4), the long-term safety benefits and reduced accidents can justify the investment.

Table 4. Bid price ranges for marking materials [91].

Materials		Paint	Epoxy	Polyurea	MMA	Thermoplastic	Tape
Price ($/ft)	4 in.	0.05–0.22	0.3–1.32	0.56–1.32	1.25	0.11–0.91	1.94–3.78
	6 in.	0.11–091	0.54–0.69	0.8	0.79–0.8	0.16–1.08	2.08–5.62

Table 5 summarizes the skid resistance test results of all pavement markings cited in Section 3 of this paper, including detailed material and glass bead information, and shows substantial variability based on marking type, binder, and test environment. Clearly, there are more lab studies than field studies. Adding anti-skid additives can greatly increase the skid resistance, especially for epoxy and TP markings. A direct comparison of the test results can lead to the following general guidelines for minimum skid resistance requirements, which ensure sufficient safety and performance across various road conditions and traffic volumes:

- 40 BPN (new markings) for low-traffic application areas;

- 45–50 BPN (new markings) for typical roads with moderate-traffic applications;
- 60 BPN or higher (new markings) for skid-prone and safety-critical locations.

Table 5. Summary of skid resistance test results from reviewed references.

Source	Marking Materials		Skid Resistance [1]	Test Environment
	Binder	Beads		
Richard (1975) [27]	Paints	Glass	30~32 (38~45)	Lab
	Cold plastic	Glass	14~38	Lab
	Extruded hot plastic	Glass	30~38	Lab
	TP tape	Glass	36~38	Lab
Anderson & Henry (1980) [2]	Paints	Glass	47~61 (28~59)	Lab
	Cold plastic	Glass	45~58 (46~57)	Lab
	Hot extruded TP	Glass	40~47 (21~39)	Lab
	Hot sprayed TP	Glass	46~63 (19~45)	Lab
Bagot (1995–1996) [28,29]	Waterborne	Silica	RFT: 0.77~0.90 (0.44~0.47)	Field
	Epoxy	Silica	0.60~0.90 (0.32~0.40)	Field
	Methacrylic resin	Silica	0.77~0.90 (0.44~0.47)	Field
	Methacrylic resin	Silica+Glass	0.42~0.52	Field
Rodin et al. (2018) [22]	Waterborne	Glass	52	Lab
	Preformed fused TP	Glass	40	Lab
	Patterned surface tape	Micro-ceramic	64	Lab
Fanijo et al. (2023) [30]	Waterborne	Glass	MTD: 0.90 (MTD: 0.62; IFI: 0.18)	Lab
	Thermoplastic	Glass	IFI: 0.19; MTD: 1.20	Lab
	MMA	Intermix glass	IFI: 0.40; MTD: 0.62	Lab
Bao et al. (2024) [31]	Waterborne	Glass+&-ceramic	MPD: 0.31–1.26; BPN: 40.0–62.5	Lab
	Preformed tape	Micro-ceramic	MPD: 1.08; BPN: 40.8	Lab
	Epoxy	Glass	MPD: 0.42–0.75; BPN: 33.8–35.5	Lab
	Polyurea	Glass+&-ceramic	MPD: 0.37–1.55; BPN: 34.0–42.5	Lab
	MMA	Glass+Corundum	MPD: 0.34, BPN: 47.5	Lab
	TP	Glass+Micro-ceramic	MPD: 0.25–0.26; BPN: 30.5–32.5	Lab
Reid et al. (1962) [32]	TP	Glass	37~58	Lab
	TP	Glass+Gritty agg.	48~90	Lab
	TP	Reflectorized	44~70	Field
Purohit et al. (2020) [33]	Preformed TP	Glass	69~73; MTD: 2.8~4.2	Field
Pasetto & Manganaro (2006) [9]	Solvent-based paints	Glass+Anti-skid granules	SRT: 41~55; MTD: 0.51~0.91	Field
Lundkvist & Isacsson (2007) [35]	TP	Glass	SRT: 0.63~0.92; MTD: 0.15~2.35	Field
Coves-Campos et al. (2018) [7]	Styrene acrylic paints	Glass±Anti-skid agg.	SRT: 49~63	Field
Burghardt et al. (2023) [8]	Paints	Glass	SRT: 44 (35)	Field
	Paints	Glass+Corundum	50~55	Field
	Paints	Andi-skid agg.	48~65	Field
Thew & Dabic (2000) [36]	Waterborne	Intermix Type C glass	46~70	Field
Naidoo & Steyn (2018) [37]	Waterborne	Glass	SRT: >50	Lab.
	Cold plastic	Glass	>50	Lab.
	TP	Glass	>50	Lab.
	Others	Glass	35–45	Lab.
Siyahi et al. (2015) [38]	Acrylic paints	Glass	41~48 (33)	Lab.
Hadizadeh et al. (2020) [39]	MMA	Silica	48~74	Lab.
Wang et al. (2023) [40]	Solvent-based paints	Glass	70 (35)	Lab.
Chen et al. (2023) [10]	Hydrophobic hot-melt paint	Glass	65	Lab.
	Traditional paint	Glass	68	Lab.
Yang (2023) [11]	High-performance paint	Glass	65	Lab.
Chiu et al. [41]	Heat-treated polyester	Intermix glass	45~68	Field
Su et al. (2021) [42]	TP (BPN ≥ 65)	Glass+Anti-skid agg.	60 or 71	Field or Lab.
	TP	Anti-skid agg.	52 (at 65 days)	Field
	Heat-treated polyester	Anti-skid agg.	95	Field

[1] Unless otherwise specified, skid resistance is in BPN, MPD is in mm, and values in brackets refer to un-beaded markings.

Note that thermoplastic and epoxy markings commonly require anti-skid additives to ensure adequate skid resistance. Moreover, the skid resistance of pavement markings may deteriorate over time due to factors such as traffic wear, environmental exposure, material quality, and substrate characteristics. This degradation increases the risk of skidding, especially at intersections and curves in wet conditions. Localized proactive measures to address these effects can enhance road safety and optimize maintenance efforts.

6. Conclusions

This paper presents a comprehensive synthesis of the global variability in materials and skid resistance requirements for pavement markings. The diverse thresholds for skid resistance, ranging from 45 BPN to over 60 BPN, reflect the need for region-specific strategies. Areas with high traffic volumes, extreme weather conditions, or safety-critical locations (such as intersections and crosswalks) may require higher skid resistance. To adapt to local conditions, highway agencies should consider factors such as road users, materials, weathers, traffic wear, and maintenance needs.

The paper also identifies a key challenge: non-uniformity in testing methodologies. Variations in testing methods can lead to inconsistent skid resistance evaluations. Standardizing test methods, such as EN 13036 SRT and ASTM E303 BPN tests, would allow for more consistent and comparable results across different regions, ensuring that pavement markings perform as expected in various environments.

Moreover, the addition of anti-skid aggregates, such as glass beads and corundum, significantly improves skid resistance, particularly in thermoplastic and epoxy markings. However, the choice of additives should be made according to local environmental factors and road conditions.

Finally, future efforts should focus on long-term performance data for pavement markings under diverse real-world environmental conditions and traffic patterns. More detailed studies on the influence of variables like temperature, traffic load, and age on skid resistance would support the development of more robust safety standards. In short, adapting to regional needs, standardizing testing methods, and further investigating the long-term impacts of marking materials will help enhance the skid resistance and safety of pavement markings globally.

Author Contributions: Conceptualization, S.L., Y.J., J.B. and H.Z.; methodology, J.B., H.Z. and S.L.; validation, S.L. and Y.J.; formal analysis, J.B., H.Z. and S.L.; investigation, J.B., H.Z. and S.L.; resources, Y.J. and S.L.; data curation, J.B. and H.Z.; writing—original draft preparation, J.B., H.Z. and S.L.; writing—S.L. and Y.J.; visualization, J.B. and H.Z.; supervision, Y.J. and S.L.; project administration, Y.J. and S.L.; funding acquisition, Y.J. and S.L. All authors have read and agreed to the published version of the manuscript.

Funding: This research was partially funded by the Joint Transportation Research Program (JTRP), grant number SPR-4646.

Data Availability Statement: The data analyzed during this study are available from the corresponding author upon reasonable request.

Acknowledgments: This work was partially funded by the Joint Transportation Research Program (JTRP) administered by Purdue University and the Indiana Department of Transportation. The authors would like to thank the Study Advisory Committee members: Tim Wells, Samy Noureldin, Dave Boruff, Subhi Bazlamit, and Bart Williamson of INDOT, Eryn Fletcher of FHWA Indiana Division, and Kenneth Ferguson of 3M for their technical guidance. Special thanks are extended to Limin Du of Shanxi Changda Traffic Facilities Co., Ltd, Yu-Min Su of National Kaohsiung University of Science and Technology, and Daehyeon Kim of Chosun University for their assistance in collecting references. The contents of this paper reflect the views of the authors, who are responsible for the facts and the accuracy of the data presented herein, and do not necessarily reflect the official views or policies of the sponsoring organization. These contents do not constitute a standard, specification, or regulation.

Conflicts of Interest: The authors declare no conflicts of interest.

References

1. FHWA. *Manual on Uniform Traffic Control Devices for Streets and Highways*, 11th ed.; FHWA: Washington, DC, USA, 2023.
2. Su, Y.; Chen, J.; Cheng, J.; Hsu, Y.; Huang, M. Rough-set based association rules toward performance of high-friction road markings. *J. Transp. Eng. Part B Pavements* **2022**, *148*, 05022001. [CrossRef]
3. *EN 1436:2018*; Road Marking Materials-Road Marking Performance for Road Users and Test Methods. European Committee for Standardization (CEN): Brussels, Belgium, 2018.

4. Carlson, P.; Park, E.; Pike, A.; Porter, R.J.; Miles, J.; Boulanger, B.; Smadi, O.; Hawkins, N.; Chalmers, S.; Darmiento, F.; et al. *Pavement Marking Demonstration Projects: State of Alaska and State of Tennessee*; No. FHWA-HRT-12-048; Texas A&M University System Transportation Institute: College Station, TX, USA, 2013.
5. FHWA. *National Standards for Traffic Control Devices; the Manual on Uniform Traffic Control Devices for Streets and Highways; Maintaining Pavement Marking Retroreflectivity*; 23 CFR Part 655; FHWA: Washington, DC, USA, 2022.
6. National Highways. *Design Manual for Roads and Bridges: Volume 8, Section 2, Part 2 TD 26/17: Inspection and Maintenance of Road Markings and Road Studs on Motorways and All-Purpose Trunk Roads*; National Traffic Operations Centre: Birmingham, UK, 2017.
7. Coves-Campos, A.; Bañón, L.; Coves-García, J.A.; Ivorra, S. In situ study of road marking durability using glass microbeads and antiskid aggregates as drop-on materials. *Coatings* **2018**, *8*, 371. [CrossRef]
8. Burghardt, T.E.; Köck, B.; Pashkevich, A.; Fasching, A. Skid resistance of road markings: Literature review and field test results. *Roads Bridges—Drog. I Mosty* **2023**, *22*, 141–165. [CrossRef]
9. Pasetto, M.; Manganaro, A. Study on the effect of surface texture saturation of road pavements with drop on road markings. In Proceedings of the 5th Pan-European Conference on Planning for Minerals and Transport Infrastructure, Sarajevo, Bosnia, 18 May 2006.
10. Chen, J.; Li, R.; Zhang, Y.; Wu, Y.; He, H. Study on the reflective principle and long-term skid resistance of a sustainable hydrophobic hot-melt marking paint. *Sustainability* **2023**, *15*, 9950. [CrossRef]
11. Yang, J.Y. Key technology and engineering application of high-performance highway color anti-skid line marking. *Commun. Sci. Technol. Heilongjiang* **2023**, *46*, 35–37. (In Chinese)
12. Anderson, D.A.; Henry, J.J. Wet-pavement friction of pavement-marking materials. *Transp. Res. Rec.* **1980**, *777*, 58–62.
13. Anderson, D.A.; Henry, J.J.; Hayhoe, G.F. Prediction and significance of wet skid resistance of pavement marking materials. *Transp. Res. Rec.* **1982**, *893*, 27–32.
14. Ouellet, J.V. Environmental hazards in motorcycle accidents. In Proceedings of the 26th Annual Meeting of the American Association for Automotive Medicine, Ottawa, ON, Canada, 4–6 October 1982; Volume 325.
15. Gates, T.J.; Hawkins, H.G. *The Use of Wider Longitudinal Pavement Markings*; Report No. 02-0024-1; Texas A&M University System Transportation Institute: College Station, TX, USA, 2002.
16. Jiang, Y. *Durability and Retro-Reflectivity of Pavement Markings*; Publication No. FHWA/IN/JTRP-2007/11; Joint Transportation Research Program, Indiana Department of Transportation and Purdue University: West Lafayette, IN, USA, 2008.
17. Xu, L.; Chen, Z.; Li, X.; Xiao, F. Performance, environmental impact and cost analysis of marking materials in pavement engineering, the-state-of-art. *J. Clean. Prod.* **2021**, *294*, 126302. [CrossRef]
18. Andrady, A.L. *Pavement Marking Materials: Assessing Environment-Friendly Performance*; NCHRP Report 392; Transportation Research Board: Washington, DC, USA, 1997.
19. Migletz, J.; Graham, J.L.; Bauer, K.M.; Harwood, D.W. Field surveys of pavement-marking retroreflectivity. *Transp. Res. Rec.* **1999**, *1657*, 71–78. [CrossRef]
20. Migletz, J.; Graham, J.L. *Long-Term Pavement Marking Practice*; NCHRP Synthesis 306; Transportation Research Board: Washington, DC, USA, 2002.
21. Lopez, C.A. *Pavement Marking Handbooks*; Taxes Department of Transportation: Austin, TX, USA, 2004.
22. Rodin, H., III; Nassiri, S.; Yekkalar, M. *Evaluation of Motorcyclists' and Bikers' Safety on Wet Pavement Markings*; PacTrans: Seattle, WA, USA, 2018.
23. Methyl Methacrylate Traffic Paint. Available online: https://mets.dot.ca.gov/aml/MethylMethacrylateTrafficPaint.php (accessed on 8 November 2024).
24. Rencheck, M.L.; Gohl, J.A.; Grennan, H.P.; Erk, K.A.; Davis, C.S. Assessing the elastic moduli of pavement marking tapes using the tape drape test. *Transp. Res. Rec.* **2021**, *2675*, 570–579. [CrossRef]
25. *AASHTO M247-13*; Standard Specification for Glass Beads Used in Pavement Markings. AASHTO: Washington, DC, USA, 2018.
26. INDOT. *Indiana Design Manual*; Indiana Department of Transportation (INDOT): Indianapolis, IN, USA, 2013.
27. Richard, C.L. *Skid Testing of Pavement Markings*; Publication TSD-177-75; Michigan Department of Transportation: Lansing, MI, USA, 1975.
28. Bagot, K. *Evaluation of Alternative Pavement Marking Materials*; Publication DOT/FAA/CT-94/119; FAA: Washington, DC, USA, 1995.
29. Bagot, K. *Follow-Up Friction Testing of Retro-Reflective Glass Beads*; Publication DOT/FAA/AR-TN96/74; FAA: Washington, DC, USA, 1996.
30. Fanijo, E.O.; Kassem, E.; Mohamed, M.; Lowry, M. Performance Evaluation of Green Pavement Markings for Bicycle Infrastructure. *J. Transp. Eng. Part B Pavements* **2023**, *149*, 04022071. [CrossRef]
31. Bao, J.; Hu, X.; Peng, C.; Duan, J.; Lin, Y.; Tao, C.; Jiang, Y.; Li, S. *Advancing INDOT's Friction Test Program for Seamless Coverage of System: Pavement Markings, Typical Aggregates, Color Surface Treatment, and Horizontal Curves*; Publication No. FHWA/IN/JTRP-2024/09; Joint Transportation Research Program, Indiana Department of Transportation and Purdue University: West Lafayette, IN, USA, 2024.

32. Reid, J.A.; Sabey, B.E.; James, J.G. White-line road-markings and light-coloured road surfacings. *J. Appl. Chem.* **1962**, *12*, 201–217. [CrossRef]
33. Purohit, K.; Rahman, M.; Price, A.; Woodside, A. Assessment of preformed 3D-thermoplastic road markings for long-term durability, skid resistance and texture functionality. In Proceedings of the 9th International Conference on Maintenance and Rehabilitation of Pavements–Mairepav9, Zurich, Switzerland, 1–3 July 2020.
34. Asdrubali, F.; Buratti, C.; Moretti, E.; D'Alessandro, F.; Schiavoni, S. Assessment of the performance of road markings in urban areas: The outcomes of the CIVITAS Renaissance Project. *Open Transp. J.* **2013**, *7*, 7–9. [CrossRef]
35. Lundkvist, S.O.; Isacsson, U. Prediction of road marking performance. *J. Transp. Eng.* **2007**, *133*, 341–346. [CrossRef]
36. Thew, C.; Dabic, T. Alkyd vs chlorinated rubber vs. waterborne roadmarking paints. In Proceedings of the Queensland Conference of the Institute of Public Works Engineering Australia, Port Douglas, Queensland, Australia, 1–6 October 2000.
37. Naidoo, S.; Steyn, W.J. Performance of thermoplastic road-marking material. *J. S. Afr. Inst. Civ. Eng.* **2018**, *60*, 9–22. [CrossRef]
38. Siyahi, A.; Kavussi, A.; MIRZA, B.A. Enhancing skid resistance of two-component road marking paint using mineral and recycled materials. *Int. J. Transp. Eng.* **2015**, *3*, 195–205.
39. Hadizadeh, E.; Pazokifard, S.; Mirabedini, S.M.; Ashrafian, H. Optimizing practical properties of MMA-based cold plastic road marking paints using mixture experimental design. *Prog. Org. Coat.* **2020**, *147*, 105784. [CrossRef]
40. Wang, D.; Yang, X.; Chu, X.; He, Y.; Fan, Z.; Xing, C.; Liu, P. A novel method for evaluating the durability and environmental pollution of road markings on asphalt pavement. *J. Clean. Prod.* **2023**, *411*, 137255. [CrossRef]
41. Chiu, J.C.; Chu, C.T.; Ho, H.W.; Huang, S.C. A Study on the Retroreflectivity of Heat-Treated Polyester Pavement Markings. *Taiwan Highw. Eng.* **2017**, *43*, 2–27. (In Chinese)
42. Su, Y.M.; Chen, J.H.; Hsu, Y.T.; Cheng, J.Y.; Chang, K.K.; Yeh, T.H.; Lai, C.H.; Huang, M.C. *A Study on Skid Resistance in Pavement Marking Materials*; National Kaohsiung University of Science and Technology: Kaohsiung, Taiwan, 2021. (In Chinese)
43. Su, Y.M.; (National Kaohsiung University of Science and Technology, Kaohsiung, Taiwan). Personal communication, 26 September 2024.
44. FDOT. *Standard Specifications for Road & Bridge Construction*; Florida Department of Transportation (FDOT): Tallahassee, FL, USA, 2024.
45. *CSJ 6147-86-001*; Special Specification 7270-High Build Paint. Texas Department of Transportation (TxDOT): Austin, TX, USA, 2024.
46. GDOT. *Special Provision Section 657–Preformed Plastic Pavement Markings*; Georgia Department of Transportation (GDOT): Atlanta, GA, USA, 2012.
47. IDOT. *Standard Specifications for Road and Bridge Construction*; Illinois Department of Transportation (IDOT): Springfield, IL, USA, 2022.
48. CDOT. *Standard Specifications for Road and Bridge Construction*; Colorado Department of Transportation (CDOT): Denver, CO, USA, 2023.
49. MnDOT. *Standard Specifications for Road and Bridge Construction*; Minnesota Department of Transportation (MnDOT): St. Paul, MN, USA, 2020.
50. NDDOT. *Standard Specifications for Road and Bridge Construction*; North Dakota Department of Transportation (NDDOT): Bismarck, ND, USA, 2024.
51. NDOT. *Standard Specifications for Road and Bridge Construction*; Nevada Department of Transportation (NDOT): Carson, NV, USA, 2014.
52. City of Knoxville. *Technical Specifications for Retroreflective Preformed Pavement Markings*; City of Knoxville: Knoxville, TN, USA, 2022.
53. City of Lawrence. *Construction Material Specifications*; City of Lawrence: Lawrence, KS, USA, 2024.
54. City of Lincoln. *Standard Specifications*; City of Lincoln: Lincoln, NE, USA, 2020.
55. Highways Agency. *Design Manual for Roads and Bridges: Volume 8: Traffic Signs and Lighting, Section 2: Traffic Signs and Road markings, Part 2: Inspection and Maintenance of Road Markings and Toad Studs on Motorways and All-Purpose Tuck Roads*; National Highways: Guildford, UK, 2020.
56. IDRRIM. *Guide de la Signalisation Horizontale*; Institut Des Routes, des Rues et des Infrastructures pour la Mobilité (IDRRIM): Paris, France, 2019.
57. Fors, C.; Johansen, T.C.; Fager, H. *Nordic Certification System for Road Marking Materials: Version 10: 2024*; Swedish National Road and Transport Research Institute: Linköping, Sweden, 2024.
58. Jeliński, Ł.; Wachnicka, J.; Jamroz, J.; Kalisz, M.; Kaźmierczak, P.; Lusa, R.; Skwierczyński, P. Testing the durability and function of road traffic management devices. *MATEC Web Conf.* **2017**, *122*, 02006. [CrossRef]
59. *NZTA M07*; Specification for Road Marking Materials. NZ Transport Agency Waka Kotahi: Wellington, NZ, USA, 2022.
60. *NZTA M20*; Specification for Long-Life Roadmarking Materials. NZ Transport Agency Waka Kotahi: Wellington, NZ, USA, 2003.
61. *AS 4049.2-2005 (R2016)*; Paints and Related Materials–Pavement Marking Materials–Thermoplastic Pavement Marking Materials–For Use with Surface Applied Glass Beads. Standards Australia: Sydney, Australia, 2016.
62. *AS 4049.4-2006 (R2016)*; Paints and Related Materials–Pavement Marking Materials–High Performance Pavement Marking Systems. Standards Australia: Sydney, Australia, 2016.
63. *GB/T 16311-2009*; Specification and Test Method for Road Traffic Markings. General Administration of Quality Supervision, Inspection and Quarantine and Standardization Administration: Beijing, China, 2010. (In Chinese)

64. *GB/T 16311 (approval draft)*; Specification and Test Method for Road Traffic Markings. General Administration of Quality Supervision, Inspection and Quarantine and Standardization Administration: Beijing, China, 2024. (In Chinese)
65. Ministry of Transportation and Communications. *Traffic Engineering Specifications–General Explanation of Some Article Revisions*; Ministry of Transportation and Communications: Taipei, Taiwan, 2021. (In Chinese)
66. *CNS 15834:2015*; Road Marking Performance for Road Users. Bureau of Standards, Metrology, and Inspection: Taipei, Taiwan, 2015. (In Chinese)
67. Tung, H. Dual Performances Road Marking's Long-Term Inspection and Cost-Effectiveness Analysis. In Proceedings of the New Construction Office, Public Works Department Expert Symposium on the Tracking Results of Skid Resistance and Retroreflectivity Dual Performance of Thermoplastic Markings, Taipei, Taiwan, 23 August 2024. (In Chinese).
68. Kummer, H.W.; Meyer, W.E. *Tentative Skid-Resistance Requirements for Main Rural Highways*; NCHRP Report 37; Highway Research Board: Washington, DC, USA, 1967.
69. Zhao, G.; Liu, L.; Li, S.; Tighe, S. Assessing pavement friction need for safe integration of autonomous vehicles into current road system. *J. Infrastruct. Syst.* **2021**, *27*, 04021007. [CrossRef]
70. AASHTO. *A Policy on Geometric Design of Highways and Streets*, 7th ed.; American AASHTO: Washington, DC, USA, 2018.
71. Li, S.; Zhu, K.; Noureldin, S. Considerations in developing a network pavement inventory friction test program for a state highway agency. *J. Test. Eval.* **2005**, *33*, 287–294. [CrossRef]
72. Carter, T.J.; Enderle, B.E.; Gambardella, C.B.; Trester, R.J. Measurement of motorcycle slide coefficients. *SAE Trans.* **1996**, *105*, 1440–1447.
73. McNally, B.F.; Bartlett, W. Motorcycle sliding coefficient of friction tests. *Accid. Reconstr. J.* **2007**, *17*, 47–49.
74. AAHTO. *Guide for the Development of Bicycle Facilities, 4th ed*; AASHTO: Washington, DC, USA, 2012.
75. Government of South Australia. *Guide to Bikeway Pavement Design Construction & Maintenance for South Australia*; Government of South Australia: Adelaide, Australia, 2015.
76. Ministry of Land, Infrastructure and Transport. *Bicycle Facilities Installation and Management Guidelines*; Ministry of Land, Infrastructure and Transport: Sejong-si, Republic of Korea, 2022. (In Korean)
77. Andalusian Department of Development, Land Management and Housing. *Design Recommendations for Bicycle Paths in Andalusia*; Andalusian Department of Development, Land Management and Housing: Seville, Spain, 2013.
78. *ASTM D2047-17*; Standard Test Method for Static Coefficient of Friction of Polish-Coated Flooring Surfaces as Measured by the James Machine. ASTM: West Conshohocken, PA, USA, 2018.
79. Yamada, Y.; Tanabe, Y.; Mise, S. Research on the required skid resistance of sidewalk pavement surfaces. *J. Jpn. Counc. Traffic Sci.* **1988**, *18*, 61–65. (In Japanese)
80. Tanaka, T.; Uchida, K. *Evaluation of Comfort and Safety in Pedestrian Sidewalk Pavements*; Annual Report of the Tokyo Metropolitan Civil Engineering Research Institute: Tokyo, Japan, 1989; pp. 15–26. (In Japanese)
81. Miyata, K.; Murai, T.; Tachima, Y.; Yamada, Y. Study on the skid resistance standards for pedestrian sidewalk pavements. *J. Jpn. Soc. Civ. Eng.* **1996**, *550*, 205–212. (In Japanese)
82. *EN13036-4:2011*; Method for Measurement of Slip/Skid Resistance of a Surface: The Pendulum Test. European Committee for Standardization (CEN): Brussels, Belgium, 2011.
83. *ASTM E303-93*; Standard Test Method for Measuring Surface Frictional Properties Using the British Pendulum Tester. ASTM: West Conshohocken, PA, USA, 2019.
84. INDOT. *Standard Specifications*; Indiana Department of Transportation (INDOT): Indianapolis, IN, USA, 2024.
85. *OPSS.PROV 710*; Construction Specification for Pavement Markings. Ontario Ministry of Transportation: Toronto, ON, Canada, 2023.
86. *BS EN 1436:2018*; Road Marking Materials-Road Marking Performance for Road Users and Test Methods. British Standards Institution (BSI): London, UK, 2018.
87. *4049.1-2005 (R2016)*; Paints and Related Materials–Pavement Marking Materials–Solvent-borne Paints–For Use with Surface Applied Glass Beads. Standards Australia: Sydney, Australia, 2016.
88. *AS 4049.3-2005 (R2016)*; Paints and Related Materials–Pavement Marking Materials–Waterborne Paint–For Use with Surface Applied Glass Beads. Standards Australia: Sydney, Australia, 2016.
89. JCASM. *Road Marking Handbook*; Japan Contractors Association of Traffic Signs and Lane Markings (JCASM): Tokyo, Japan, 2018. (In Japanese)
90. *KS M 6080: 2022*; Paints for Road Markings. Industrial Standards Council: Seoul, Republic of Korea, 2022. (In Korean)
91. Pike, A.M.; Bommanayakanahalli, B. Development of a pavement marking life cycle cost tool. *Transp. Res. Rec.* **2018**, *2672*, 148–157. [CrossRef]

lubricants

MDPI

Article

Analysis of the Wear Mechanism and the Influence of the Chemical Composition and Repair Welds of the Pig Iron Wagon Wheels

Janusz Krawczyk, Łukasz Frocisz and Piotr Matusiewicz *

Faculty of Metals Engineering and Industrial Computer Science, AGH University of Krakow, A. Mickiewicza 30, 30-059 Krakow, Poland; jkrawcz@agh.edu.pl (J.K.); lfrocisz@agh.edu.pl (Ł.F.)
* Correspondence: matus@agh.edu.pl

Abstract: The aim of the present study was to develop a concept for the inverse analysis of wear mechanisms in cast steel wheels of a pig iron wagon after long-term operation. Samples were taken from the flange and the tread edge area of fourteen wheels. The impact of wheel parameters and repair methods on their wear was assessed. An analysis was carried out to determine whether welds were made as part of casting correction or as repair welds. Changes in the microstructure of the weld area, the heat-affected zone, and the parent material resulting from operation were determined. The main wear mechanism in the area of the welds and the parent material is the plastic flow of the material resulting from high unit pressures. The hardness of the material is found to be contingent upon its chemical composition, the microstructural components, and the degree of plastic deformation resulting from wear (it has been established that increasing alloying results in increased hardness; a comparable effect is observed in the formation of non-equilibrium structures (bainite)). The increase in hardness is attributed to strain hardening, a consequence of exploitation. Research and analytical methods have been developed to differentiate the results of repair processes for wear effects in a highly loaded friction node with non-stationary lubrication conditions from repair processes applied to castings of large structural components.

Keywords: pig iron wagon; cast iron; pad weld; wear; regeneration

1. Introduction

The smelting of raw iron from ore is typically accomplished through the utilization of the blast furnace process [1–3]. The resultant pig iron is then transported via rail to the bottling stations, mostly using the ladles for this purpose. The smelted ore is transported by rail using specialized wagons known as pig iron wagons [4].

The design of special railway platforms employed within the metallurgical industry is mainly based on a designed wheel set, comprising an axle and two wheels, the function of which is to roll on the rails. The main tasks of the wheel set include [1,5]:

- allowing the vehicle to move on the track;
- transferring the weight of the loaded vehicle;
- transferring the driving and braking forces, as well as dynamic loads caused by track unevenness.

Monolithic wheels may have indirect or direct contact with the rail. In railway transport, two types of wheels can be distinguished (Figure 1) [1,5]:

- monoblock wheels;
- tired wheels, which include a bare wheel and the tire around the outside.

Figure 1. Comparison of bare wheel (**a**) and tyred wheel (**b**) [6].

The utilization of a monoblock wheel, owing to its monolithic nature, is significantly economical than a tyred wheel. This is primarily attributed to its enhanced durability and reduced manufacturing expenses [1,6].

1.1. Monoblock Wheel Construction

The monoblock wheels' monolithic nature simplifies their construction and production. The manufacturing process involves machining a monoblock to obtain the designed shape [1,5,6]. The wheel's design comprises a flange and a tread that are in direct contact with rails, and a web (Figure 2).

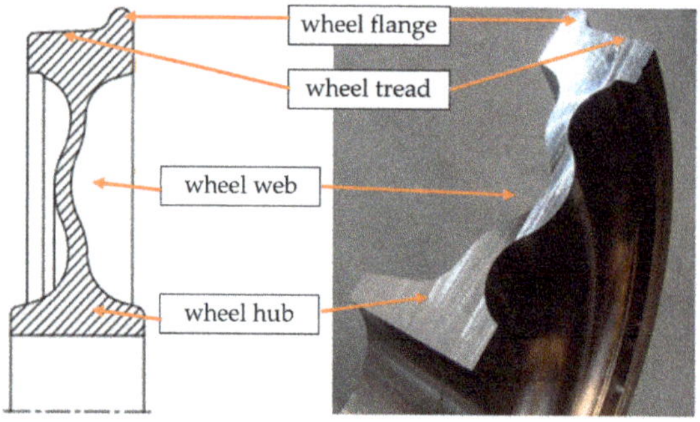

Figure 2. Cross-section of a monoblock wheel [6].

Wheel sets used in the railway industry represent critical components of rail vehicles, significantly influencing vehicle mobility and transport safety during movement. This is attributable to the impact of static and dynamic forces on the axles associated with the wheels. The utilization of these wheel sets under continuous conditions, whilst the vehicle is in motion and bearing substantial loads, including those stemming from track irregularities, results in the tribological deterioration of the surface layers of the working elements [7–10]. The decision to repair road wheels rather than replace them has been prompted by the emergence of damage in the form of large gaps and cracks. Furthermore, the continuous increase in the cost of the materials required for such repairs has led to a shift in focus towards the regeneration of existing components. This approach is regarded as a significant ecological measure, as it facilitates the reuse of parts, thereby reducing the environmental impact associated with the production of new materials [1]. One method of enhancing the durability of elements that are in direct contact with railway rails is the process of welding [11–13]. In order to ensure the effective and responsible execution of welding operations on critical elements, a set of standards has been developed. These standards, to which service companies are expected to comply, form a framework for the responsible practice of welding [1,5,14,15].

1.2. Operation of a Railway Wheel

Railway wheels are subject to a multitude of operating factors, including static loads, dynamic loads, and impact loads. These phenomena are the result of external forces, internal stresses, corrosion, thermal and mechanical phenomena, or friction. Due to the wide range of tasks assigned to wheel sets, damage occurs in the form of discontinuities or wear marks [1,16,17]. The most common damages include fatigue cracks that appear in various places on the flange and tread. The following areas are of particular significance in terms of crack formation (Figure 3):

(a) rolling surface (6–17%);
(b) outer edge (40–60%);
(c) flange (2–10%);
(d) marks on the outer side surface (3–10%);
(e) marks on the inner side surface (2–5%);
(f) defective attachment surface (10–35%).

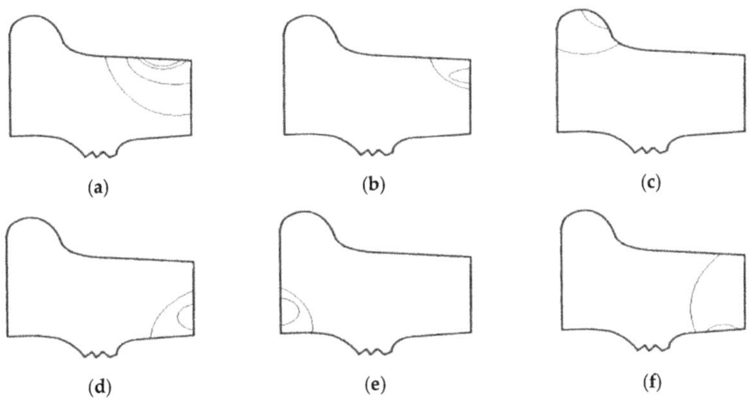

Figure 3. Location of fatigue crack formation in the wheel.

The deterioration of the monoblock wheel is contingent on the dimensions and category of the load during transportation, with the utilization of load levels that exceed the established limits resulting in a reduction in the operational longevity of the wheel set [18,19]. The prolonged use of monoblock wheels frequently results in surface destruction and the formation of cracks. The damage to the wheelset Maglio et al. conducted a study to examine the impact of railway wheel tread damage on rail wheel impact loads and the durability of wheelsets. They investigated the correlation between the magnitude of the impact load and various operational parameters and presented a methodology for predicting the effects of three-dimensional scanned wheel tread degradation on the durability of railway wheels, axles, and bearings [20].

1.3. Welding Repair Method of Cast Wheels

In order to avoid the replacement of the entire wheel, which would require a significant financial investment, the process of regeneration is performed in accordance with the [15].

This standard concerns the requirements and tests of steel castings for use in the railway industry [1,15]. The repair process may be carried out using welding or other methods agreed between the service ordering party and the contractor [15,19].

Welding is widely used in industrial processes that play a crucial role in prolonging the lifetime of the worn-out components or protecting machine parts from wear. Various welding techniques such as plasma transfer arc welding, gas metal arc welding, laser beam welding, and submerged arc welding are employed for overlay welding [11,19].

During the 1990s, the Paton Electric Welding Institute undertook extensive research on the welding and repair technology of railway wheels, which is still in use today [18,19]. The Welding Institute (TWI), as part of the AURORA project, researched the use of submerged arc welding technology for the repair of railway wheels. The primary objective of this research was to ensure stability and to minimize defects by welding [21]. Goo et al. restored the flange of the wheel, by overlay welding using the CO_2 arc welding technique and evaluated the mechanical and wear characteristics of the restored railway wheel [18]. Coo and Lee conducted a study to repair a railway wheel by welding using the submerged arc welding technique and analyzed the changes in mechanical properties due to welding [22]. Byung-Ryeol et al. undertook a study of a welding repair method that minimizes the wheel cutting thickness and extends the service life. This was achieved by locally welding and cutting the wheel flange, which is a major wear area. It is anticipated that the implementation of the welding repair method, employing submerged arc welding to achieve a bainitic microstructure in the weld metal, will have the effect of extending the service life by more than 1.5 times [19].

The process of wheel regeneration by welding is accomplished through the utilization of arc welding techniques, incorporating the application of a filler material that exhibits a chemical composition analogous to that of the wheel undergoing regeneration. Alternatively, the use of a dissimilar material, characterized by superior mechanical properties in comparison to the wear mechanisms that are the catalyst for regeneration, may be employed [12,13,18,22–24]. It is imperative that the technology employed for regeneration is subject to rigorous verification and precise adjustment. This is due to the inherent differences in microstructure that can be observed between the weld deposit, the heat-affected zone (HAZ), and the parent material. Furthermore, the potential for various welding inconsistencies, such as the formation of cracks or sticking, must be taken into consideration [25–27].

1.4. Cast Steel Used for Monoblock Wheels

In the case of monoblock wheels, it is essential to ensure not only that adequate levels of strength and wear resistance are attained but also that the requisite levels of machinability and weldability are met [9,28]. It is evident that these properties enable the acquisition of the final shape, the appropriate roughness, and subsequent repair processes [10,29]. The use of high-strength steel is not feasible as a solution to the identified requirements, primarily due to its proclivity for cracking, a consequence of both welding and the heat-deformation cycle. The manufacturing of the final wheel shape from steel is accompanied by substantial expenses stemming from material wastage during the machining process [1]. The previously mentioned requirements for road wheels imply that the most frequently used material for these components is cast steel. Casting facilitates the attainment of a wheel shape that is as close as possible to the final shape while minimizing machining and welding processes. Monoblock castings are also characterized by uniform properties in all directions, a quality that is highly desirable in these components [1]. The significant role of monoblock wheels in the context of railway transport precludes the use of all types of cast steel. Consequently, standards have been devised to present appropriate guidelines for the materials used in the manufacture of these components [1,30].

In this study, a detailed analysis was conducted on fourteen monoblock wheels that had been withdrawn from service following extended operation and potential subsequent repairs. The primary reason for this withdrawal was the identification of significant changes in the geometry of the wheel tread edge, which rendered further utilization of the wheels unfeasible. Access to reliable information regarding the manufacturing process of the wheels and their subsequent operation history was limited. The objective of the research was to develop a concept for the inverse analysis of wear mechanisms in cast steel wheels of a pig iron wagon after long-term operation, with a focus on assessing the impact of wheel parameters and repair methods on their wear. This is of significant importance for the issue of failure analysis, leading to the design of new materials used for castings in highly loaded rolling systems in changing weather conditions.

2. Materials and Methods

The material for the tests was obtained from fourteen monoblock cast steel wheels of the pig iron wagons, which exhibited signs of wear due to previous service. Samples were collected from the wheel flange and from the wheel tread edge area, i.e., the area that was exposed to the most severe working conditions. Figure 4 shows photographs of the monoblock wheels with the marked areas where the samples were taken. As part of the work, the following investigations were carried out:

- macroscopic observations;
- microstructural analysis;
- chemical composition analysis;
- hardness measurements.

The samples were embedded in resin, ground, and polished so that the surface was even, smooth, and free of scratches. In order to reveal the microstructure and potential banding, etching with 2% nital was carried out.

The macroscopic analysis was based on the observation of the geometry of the given samples in the cross-section of etched metallographic sections. ImageJ was used to assess the volume fraction of areas modified by post-casting processing. Additionally, measurements were taken of the primary austenite grain cross-sectional area, and the ferrite volume fraction was made.

A range of degrees of etching allowed for the preliminary identification of characteristic areas, including welding: pad weld, heat-affected zone, and parent material. Additionally, at the locations previously defined as deformed, three qualitative types of degree of deformation were distinguished: small, medium, and large. Microscopic documentation, hardness measurements, and chemical composition analysis were carried out on the characteristic areas of the samples examined. The documentation of the microstructure was made on a Phenom XL scanning electron microscope (Pik-Instruments, Piaseczno, Poland) and an Axiovert 200 MAT light microscope (Carl Zeiss, Chrzanów, Poland).

The chemical composition of the samples was analyzed using a FoundryMaster WAS spectrometer (Hitachi, Wellesweg, Germany) and a Phenom XL scanning electron microscope. Hardness measurements were performed using a TUKON 2500 hardness tester (WILSON-HARDNESS, Spectro–Lab, Warszawa, Poland). The Vickers method was used with a load of 294 N (HV30) and a test time of 15 s.

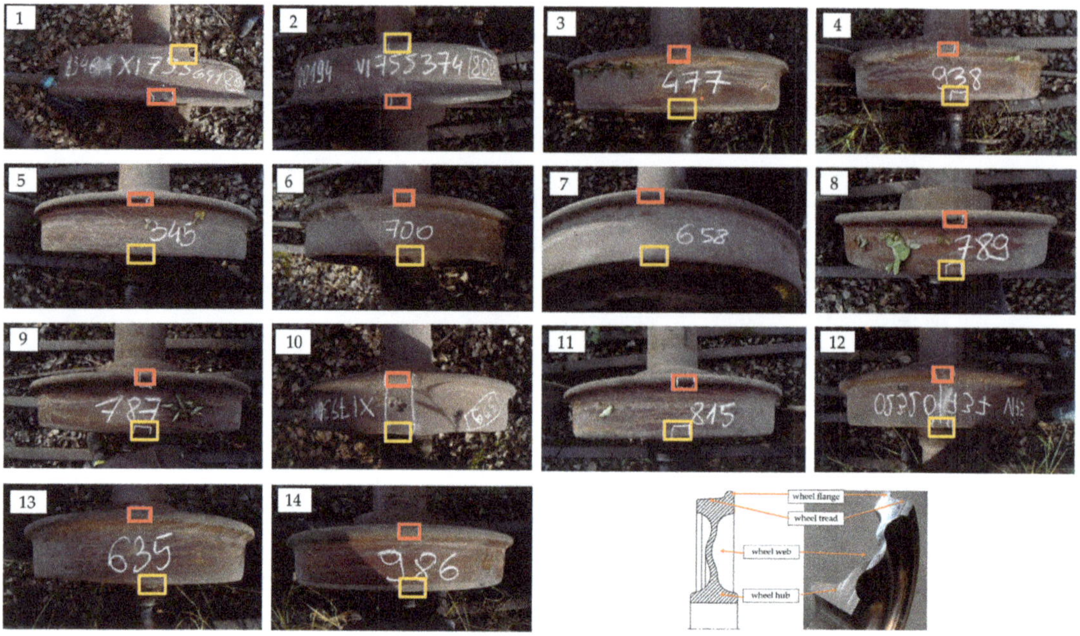

Figure 4. The tested wheels with the marked areas of sample collection: red—wheel flange, yellow—wheel tread edge area. (**1–14**): wheel number.

3. Results

3.1. Analysis of the Samples Taken from the Wheel Flange

3.1.1. Macroscopic Assessment of the Fraction of Areas Modified by Post Foundry Processing

Figure 5 shows macroscopic images of the samples cross sections with marked areas for chemical composition analysis (yellow line) and hardness measurements (red line). These macroscopic images enable an assessment of the volume fraction of areas modified as a result of post foundry processing of the tested samples. It can be noticed that samples 1, 2, and 10 exhibit the highest surface fraction of modified areas, while the macrostructure of samples 6 and 7 reveals an absence of welding treatments.

The assessment of the parent material of the cast pig iron wagon wheels was carried out on the basis of microstructural observations and hardness measurements within the area delineated by the red line (Figure 5). For this area, the chemical composition was analyzed using two methods (spark spectrometer and EDS). In order to verify the genesis of the microstructural changes, a control analysis of the chemical composition was performed in the areas limited by the yellow line.

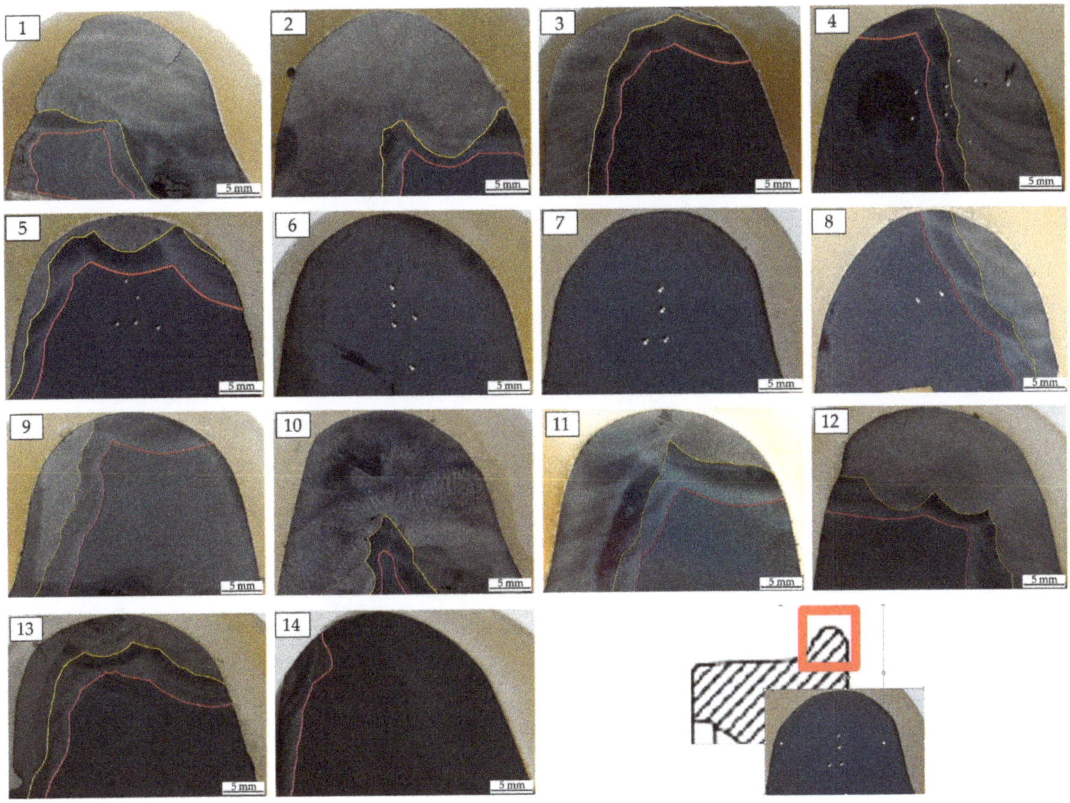

Figure 5. Macro images of etched metallographic cross sections of the wheels flange with marked hardness test areas in the parent material (red line) and fusion line (yellow line). (**1–14**): wheel number.

3.1.2. Chemical Composition Analysis

The average results of the chemical composition analysis of the parent material carried out with the spark spectrometer are presented in Table 1. It can be stated that the cast steel used is non-alloy cast steel with increased chromium content and carbon content, which is indicative of a hypoeutectoid cast steel. These cast steels can be classified within the range of materials from C45 through C50 to C55 (according to the UE classification system for unalloyed steels [31,32]). Nevertheless, based on the chemical composition results (Table 1) and the verification of areas with a modified structural character, three known steel grades were selected. The material in these areas can be classified as 100CrMn6 (in most cases), 38MnSi4, and in one case, P355NL1. It should be noted that deficiencies in the production process of cast steel monoblock wheels can be filled through the welding process. The described macrostructure of the samples is most likely the result of repair processes used in the production of cast steel monoblock wheels [22–24].

The chemical composition of the matrix was verified using a scanning electron microscope. The results of this analysis are presented in Table 2. It should be noted that the chemical composition assessment using a scanning electron microscope did not take into account the content of carbon and other microadditives and impurities. Consequently, the results indicate an average higher content of silicon, manganese, and chromium compared to the analysis performed using a spark spectrometer. However, it should be noted that this will not have a significant effect on the comparative analysis of alloy element content but rather will only affect the absolute amount. When comparing the results of chemical composition analyses in this way, there is not always a correlation between the contents of individual elements. Nevertheless, it can be hypothesized that the assessment of alloy elements performed using a scanning electron microscope more closely corresponds to the qualitative differences in the chemical composition of the matrix of the tested cast steels.

Table 1. Chemical composition of the parent material (Foundry-Master WAS spectrometer).

Wheel Number	Chemical Composition (wt. %)											
	Fe	C	Si	Mn	P	S	Cr	Mo	Ni	Al	Co	Cu
1	98.0	0.56	0.47	0.44	0.008	0.015	0.23	0.006	0.16	0.003	0.006	0.08
2	97.9	0.49	0.31	0.89	0.015	0.018	0.12	0.005	0.06	0.002	0.005	0.15
3	98.0	0.55	0.45	0.44	0.010	0.016	0.22	0.005	0.14	0.003	0.004	0.08
4	98.2	0.52	0.33	0.65	0.010	0.014	0.09	0.005	0.05	0.001	0.003	0.10
5	98.1	0.53	0.46	0.43	0.007	0.015	0.21	0.007	0.14	0.003	0.004	0.08
6	98.2	0.45	0.40	0.64	0.013	0.016	0.10	0.005	0.05	0.003	0.003	0.10
7	98.3	0.51	0.31	0.57	0.008	0.014	0.08	0.005	0.05	0.002	0.003	0.10
8	98.2	0.53	0.33	0.65	0.007	0.011	0.08	0.005	0.04	0.003	0.003	0.11
9	97.7	0.51	0.43	0.57	0.007	0.010	0.25	0.039	0.35	0.003	0.005	0.07
10	98.0	0.53	0.47	0.44	0.009	0.019	0.22	0.009	0.13	0.006	0.004	0.08
11	97.8	0.51	0.36	0.54	0.006	0.011	0.25	0.041	0.34	0.002	0.005	0.07
12	97.8	0.52	0.36	0.55	0.007	0.011	0.25	0.039	0.35	0.001	0.005	0.07
13	97.8	0.47	0.30	0.60	0.011	0.018	0.28	0.005	0.26	0.001	0.006	0.14
14	98.1	0.54	0.36	0.66	0.011	0.015	0.08	0.005	0.04	0.003	0.003	0.10

Table 2. Chemical composition of the parent material (Scanning Electron Microscope Phenom XL).

Wheel Number	Chemical Composition (wt. %)			
	Fe	Si	Mn	Cr
1	98.20	0.77	0.69	0.34
2	97.75	0.90	1.18	0.18
3	98.40	0.69	0.64	0.26
4	98.39	0.62	0.83	0.15
5	98.46	0.83	0.52	0.19
6	98.59	0.59	0.71	0.11
7	98.29	0.59	0.98	0.14
8	97.41	0.61	1.48	0.49
9	98.22	0.74	0.77	0.28
10	98.08	0.94	0.65	0.33
11	98.87	0.47	0.48	0.18
12	98.09	0.65	0.94	0.32
13	97.92	0.54	1.06	0.48
14	97.82	0.81	1.07	0.29

3.1.3. Microstructural Analysis

The microstructure of the parent material of the cast monoblock wheel tested, documented using a light microscope, is shown in Figure 6. The analyzed microstructure corresponds to a hypoeutectoid cast steel. A network of ferrite precipitates is visible along the grain boundaries of the former primary austenite. In some cases, some of the ferrite precipitations are similar to the Widmanstätten structure, indicating accelerated cooling in the temperature range of ferrite precipitation from austenite (Figure 6(1,6,12)). In certain cases (Figure 6(14)), the morphological characteristic of the matrix clearly suggests the presence of pearlitic structure. The characteristic lamellar arrangement of ferrite and cementite for the eutectoid in iron alloys is visible, but it was often difficult to observe lamellar patterns in most cases.

In order to gain a more accurate visualization of the matrix of the study materials, it was decided that observations would also be performed using a scanning electron microscopy (SEM). The matrix microstructure of the parent material in the tested pig iron wagon wheels exhibited significant differences in some cases. There are cases where the matrix is more similar to bainitic structures (Figure 7(2,3,10,11)). These findings indicate that the wheels may have undergone accelerated cooling during the crystallization or heat treatment process.

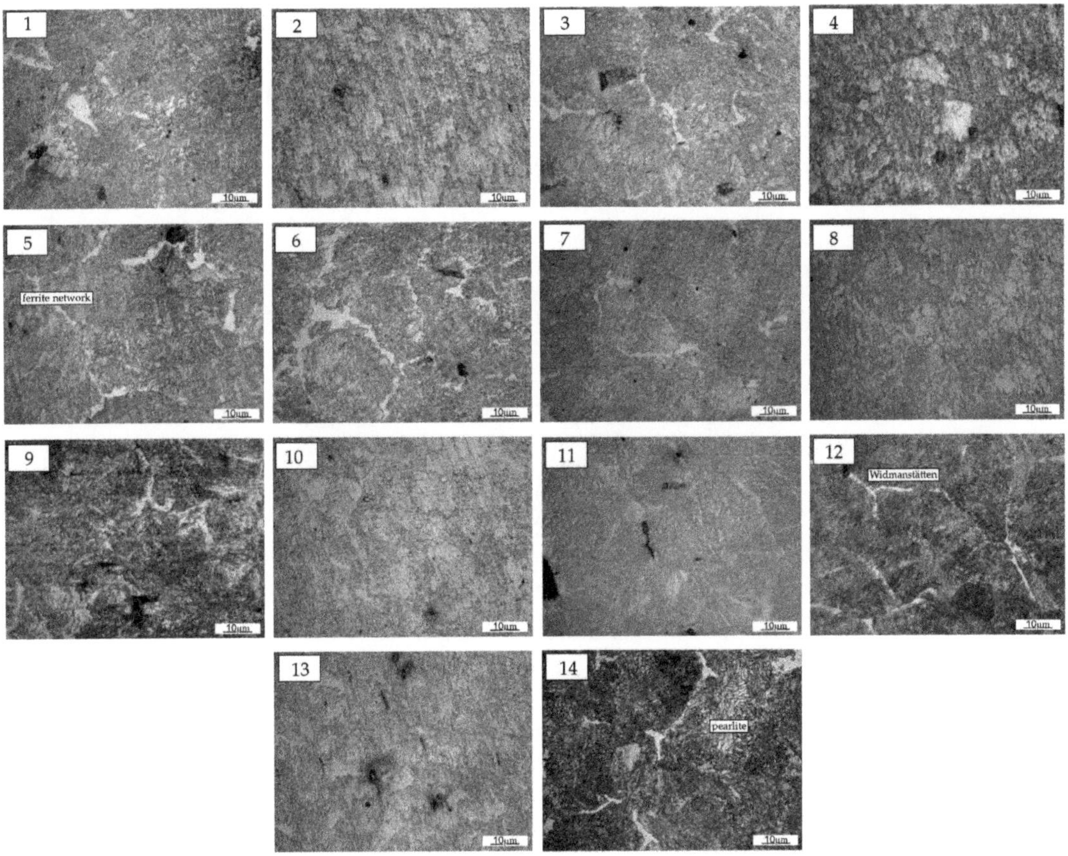

Figure 6. Microstructure of the cast steel parent material used for the production of monoblock wheels. (**1–14**): wheel number.

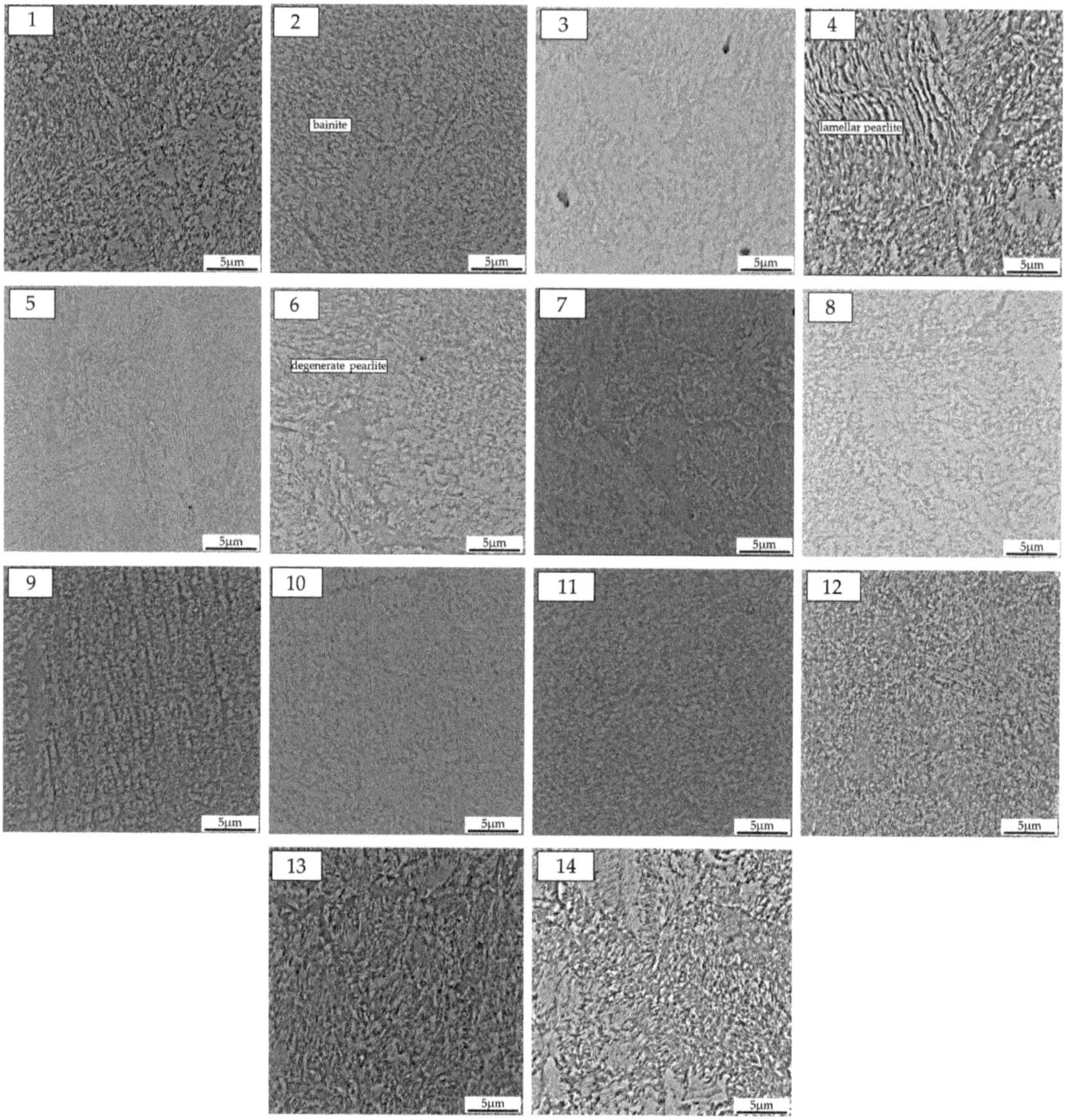

Figure 7. Microstructure of the parent material of monoblock wheels, SEM. (**1–14**): wheel number.

3.1.4. Quantitative Evaluation of Macro and Microstructural Characteristics

Further insight was gained through calculations of the fraction of modified areas (post-casting), the area of primary austenite grains, and the ferrite volume fraction; the results of the tests are presented in Table 3. Hardness measurements were made in the parent material, with the obtained results falling within the range of 257 to 297 HV30 (Table 3). It should be noted that assuming the hardness of railway rails to be approximately 300 HB (303HV30), the materials used for the pig iron wagon wheels should exhibit a marginally lower hardness. Therefore, it can be assumed that the hardnesses of the tested wheels are in alignment with the assumed requirements [8].

Table 3. Results of quantitative analysis of macro and microstructure morphological features.

Wheel Number	Hardness, HV30	Post Foundry Modified Areas Volume Fraction, %	Mean Area of Primary Austenite Grains, μm^2	Ferrite Volume Fraction, %
1	278	88	528	5.61
2	287	78	894	1.36
3	280	31	515	5.91
4	284	40	945	- *
5	292	12	465	3.48
6	257	0	481	7.42
7	259	0	645	3.18
8	293	19	805	2.88
9	297	17	594	6.52
10	287	92	746	1.82
11	288	48	847	3.03
12	297	67	1564	2.12
13	263	35	1478	1.52
14	290	0	1251	6.82

* The method of focusing sample 4 did not allow for ferrite fraction assessment—sample 4 was excluded from the analysis.

3.2. *Analysis of Samples Collected from the Wheel Tread Edge Area*

3.2.1. Macroscopic Evaluation of the Prevalence of Typical Welding Zones

Figure 8 presents macroscopic images of fourteen metallographic sections with the marked working surface (WS). Differences in the degree of etching of the samples allowed a preliminary identification of the weld zones, i.e., the padding weld (W), the heat-affected zone (HAZ), and the parent material (PM). Subsequently, these areas were later tested for chemical composition and hardness. A visual evaluation of the metallographic sections indicates that samples taken from wheels No. 1, 2, 4, 10, and 11 exhibit the highest volume fraction of areas considered to be padding welds, while samples from wheels No. 6 and 7 may indicate the absence of welding.

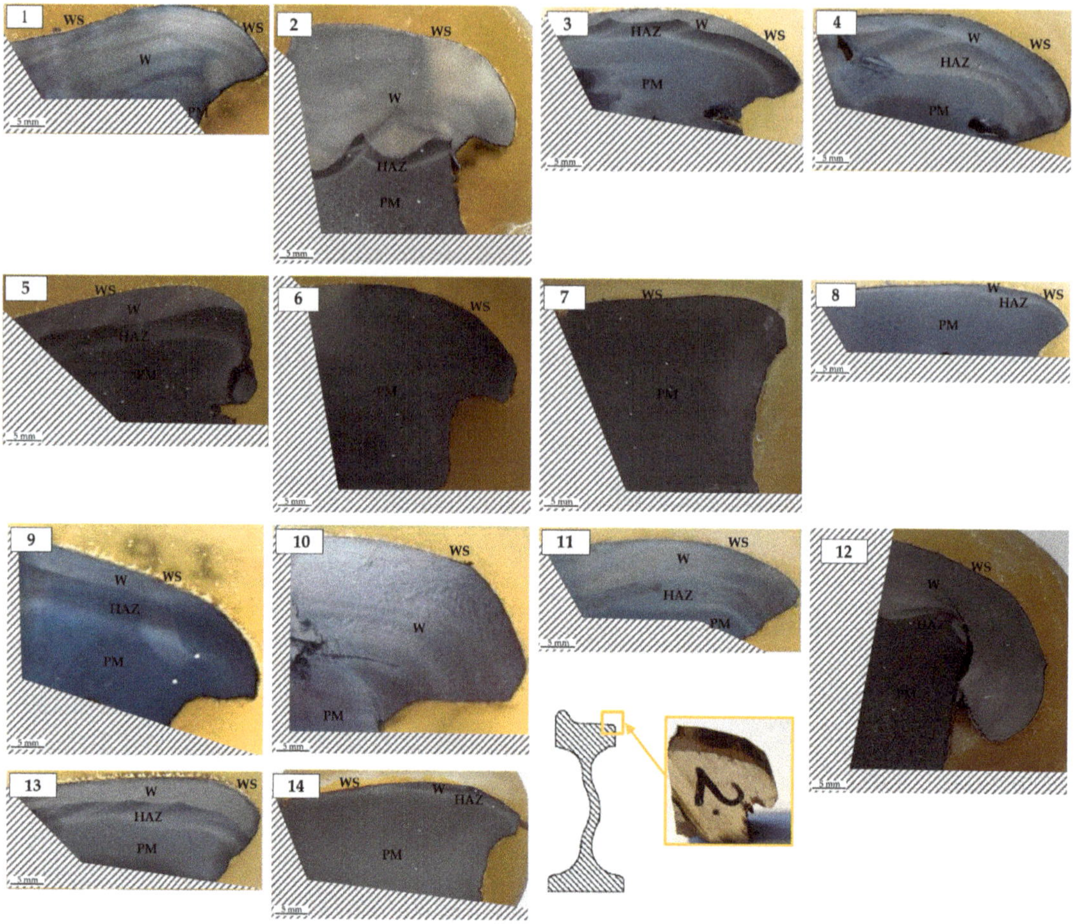

Figure 8. Macro image of etched metallographic sections with marked areas: W—weld, HAZ—heat-affected zone, PR—parent material, WS—working surface. (**1–14**): number of the wheel.

3.2.2. Chemical Composition Test Results

The chemical composition of the parent material of the undeformed areas of the wheel (i.e., the flange) was obtained using a spark spectrometer (Table 1). The results of this analysis determined that the monoblock wheels were made of hypoeutectoid unalloyed cast steel [29,30]. The chemical composition of samples taken from the wheel tread edge area was examined using an EDS (Energy Dispersive Spectroscopy) analysis in a scanning electron microscope, and the obtained results were presented in the Table 4 and the form of graphs (Figure 9) showing the contents of the given elements in the adopted padding weld and parent material. Analyzing Figure 9 reveals that, in most of the samples, the content of alloying elements in the adopted padding weld is higher than in the parent material. This finding suggests the possibility of the use of additional material during the welding process.

Figure 9. Contents of manganese (**a**), chromium (**b**), and silicon (**c**) in the areas adopted in accordance with Section 3.2.1: weld (W), parent material (PM).

Table 4. Chemical composition of the parent material (PM) and the weld (W) in samples taken from the wheel tread edge area (Scanning Electron Microscope Phenom XL).

Wheel Number	Mn (wt. %)		Si (wt. %)		Cr (wt. %)	
	Parent Material	Weld	Parent Material	Weld	Parent Material	Weld
1	1.23	1.73	0.62	0.85	0.42	0.77
2	0.90	1.48	0.68	0.81	0.11	0.39

Table 4. *Cont.*

Wheel Number	Mn (wt. %)		Si (wt. %)		Cr (wt. %)	
	Parent Material	Weld	Parent Material	Weld	Parent Material	Weld
3	0.81	1.91	0.65	1.11	0.33	0.70
4	1.04	2.14	0.62	1.46	0.21	0.71
5	0.50	2.02	0.75	1.42	0.17	0.91
6	1.09	-	0.77	-	0.28	-
7	1.14	-	0.81	-	0.30	-
8	1.13	1.60	0.49	1.13	0.26	0.76
9	0.73	2.04	0.80	1.34	0.26	0.75
10	0.60	1.52	0.79	0.71	0.21	0.63
11	1.08	2.46	1.77	2.65	0.55	1.02
12	0.87	1.48	0.50	0.56	0.33	0.55
13	1.03	2.38	1.01	2.66	0.43	1.02
14	1.07	0.97	1.30	0.93	0.00	0.15

3.2.3. Microstructural Analysis

Microstructural documentation was prepared with a light microscope for the zones adopted in Section 3.2.1 and was presented in Figure 10.

In the case of the materials tested, the microstructure of the material in the surface area is characterized by a distinct igneous morphology. The character of the microstructure of the surface varies for the different monoblock wheels tested. It is possible to observe samples characterized by a significantly larger needle size (Figure 10(4c)) and materials with a significantly finer microstructure (Figure 10(11c)). The observed differences are also significantly influenced by the plastic deformation of the material about its service life. This is particularly evident in the case of wheel 14 (Figure 10(14c)).

The microstructure of the parent material and the heat-affected zone corresponds to the pearlitic-ferritic structure (e.g., Figure 10(4a,b)). At the grain boundaries of former austenite, a lattice morphology of ferrite precipitates is also observed (e.g., Figure 10(14a)). In the regions where the padding weld is identified, the typical for this process bainitic feathery structure (e.g., Figure 10(11c)) and the bainitic structure of acicular nature (e.g., Figure 10(4c)) are observed. In the weld area of sample 12, in addition to bainite, a small fraction of fine pearlite can be seen (Figure 10(12c)). The initial interpretation of the areas of sample 14 as characterized by the presence of a padding weld is thus rather attributed to differences in the intensity of plastic deformation and thermal impact resulting from friction as a result of operation. In the weld area, a columnar structure with marked interdendritic spaces and a strongly deformed columnar structure, characteristic of the crystallization process, is visible. In places of strong deformation, a tendency for grains to elongate in the direction of plastic flow of the material was also observed (Figure 10(4c)).

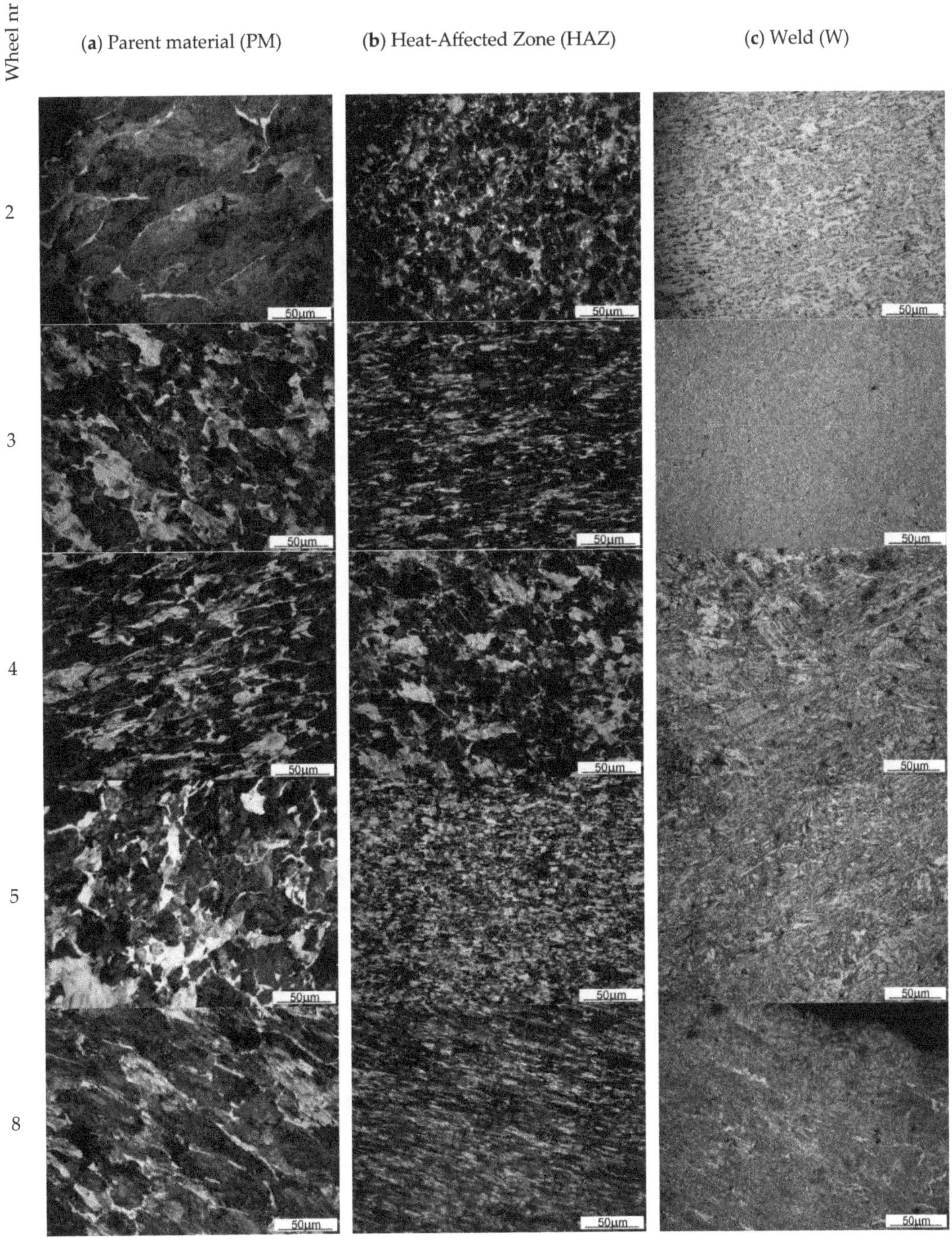

Figure 10. *Cont.*

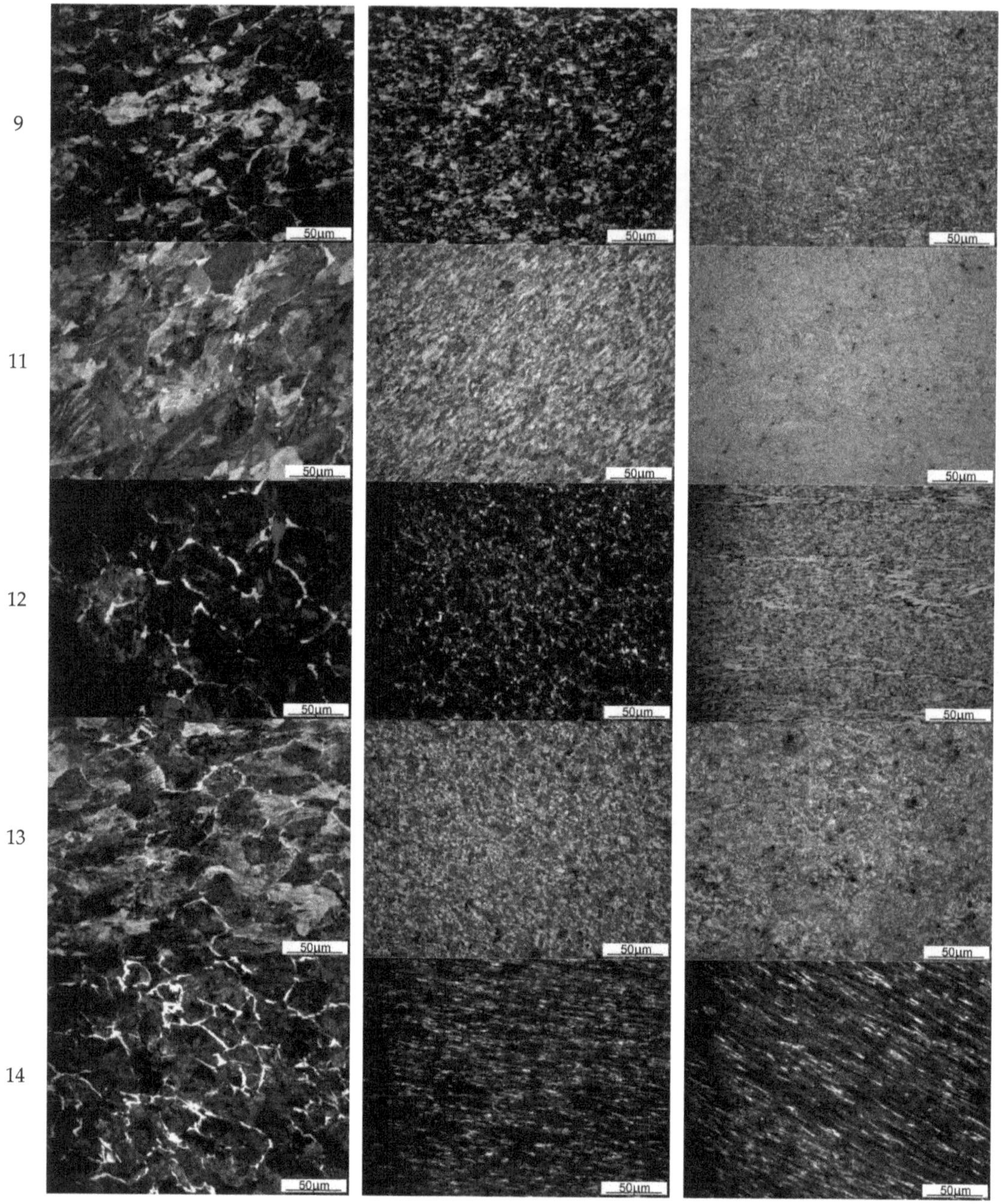

Figure 10. Microstructure of selected wheels: parent material (**a**), heat-affected zone (**b**), and weld (**c**).

A comparison of the grain size in the parent material and in the heat-affected zone reveals that, in the area created after the welding process, there is a smaller grain size. This phenomenon may be the result of recrystallization of the structure. The prepared documentation clearly shows that the degree of plastic deformation of the material decreases as it moves from the working surface into the volume of the bare wheel. This indicates that the process of plastic material flowing towards the free (side) surface of the wheel contributes significantly to the wear of bare wheels. Therefore, the properties of the parent material and the padding weld should be shaped to strengthen the material in order to resist the loss of stability to a greater extent than resistance to abrasive wear, and the rheological parameters of the padding should be considered correctly.

3.2.4. Hardness Measurement

Hardness measurements were made for each area specified in Section 3.2.1. i.e., the weld (W), the heat-affected zone (HAZ), and the parent material (PM). Analysis of the results obtained (Table 5) shows that the hardness for the assumed areas is within the following ranges: 240–430 HV30 for the weld; 230–400 HV30 for the heat-affected zone; 260–320 HV30 for the parent material.

Table 5. Hardness measurement results (HV) for parent material, heat-affected zone, and weld.

Wheel Number	Measurement Number	Weld		Heat-Affected Zone		Parent Material	
		Hardness	Mean Hardness	Hardness	Mean Hardness	Hardness	Mean Hardness
1	1	310	310		-	263	270
	2	310				283	
2	1	344	370	395	400	329	310
	2	396		407		284	
3	1	230	240	235	230	315	300
	2	250		231		277	
4	1	377	380	328	360	332	320
	2	390		400		300	
5	1	380	390	354	360	308	310
	2	405		359		315	
6	1	-	-	-	-	298	270
	2	-		-		251	
7	1	-	-	-	-	312	295
	2	-		-		263	
8	1	300	332	-	-	342	320
	2	295		-		299	
9	1	301	347	332	340	293	270
	2	425		347		238	
10	1	297	-	-	-	312	395
	2	289		-		277	
11	1	389	430	330	340	302	320
	2	465		360		337	
12	1	305	310	300	310	270	260
	2	306		318		254	
13	1	376	370	317	310	304	290
	2	370		296		278	
14	1	265	280	280	300	274	280
	2	298		311		280	

For most samples, hardness decreases from the weld through the heat-affected zone to the parent material. The values obtained for the areas created by the welding process, which reach over 350 HV, may indicate that the non-diffusion transformation of austenite to bainite has occurred during crystallization. Strain hardening due to wear can also be responsible for increased hardness in the weld area.

4. Discussion

4.1. Effect of Chemical Composition on Microstructure and Hardness

Figure 11 shows the effect of chemical composition on hardness. Utilizing a spark spectrometer, the results presented in Figure 11a,b were obtained, while the results depicted in Figure 11c–e were obtained through the implementation of EDS analysis. The observations indicate that the hardness of the material exhibits an increase in correlation with the elevated carbon content. Therefore, the volume fraction of ferrite should be expected to decrease as the carbon content increases, but the results show an inverse relationship. An increase in hardness and a decrease in the volume fraction of ferrite with increasing content of alloying elements such as Cr, Mn, and Si would be expected. Only the effect of manganese on hardness is not consistent with the above assumption from the results presented.

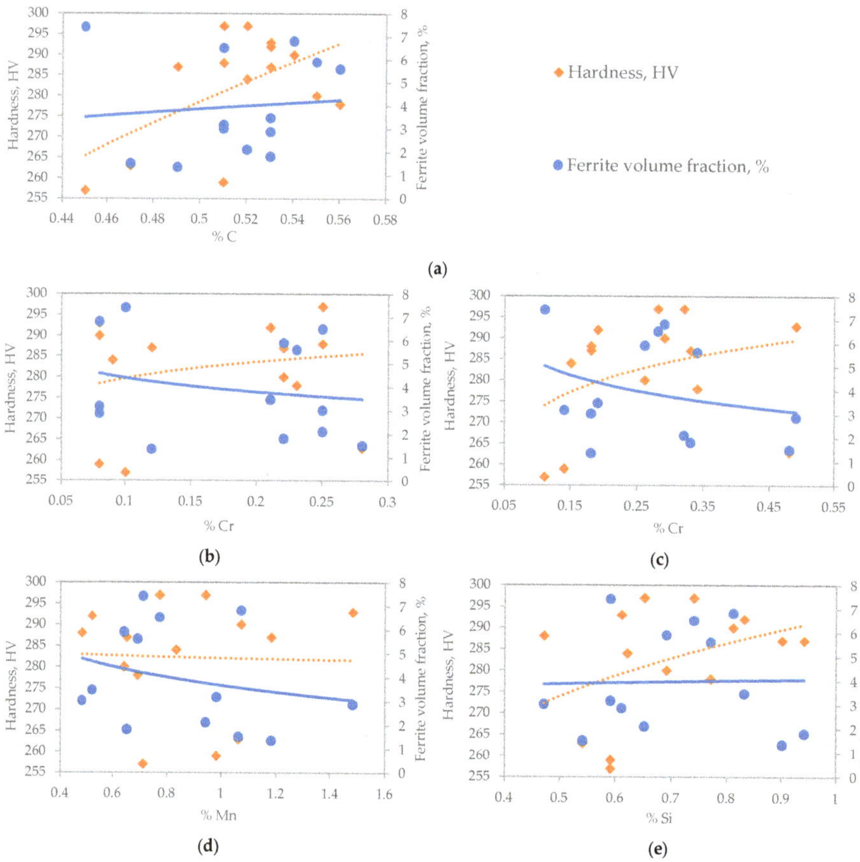

Figure 11. Dependence of hardness and ferrite volume fraction on: carbon content obtained on spark spectrometer (**a**), chromium content obtained on spark spectrometer (**b**), chromium content from EDS analysis (**c**), manganese content from EDS analysis (**d**), silicon content from EDS analysis (**e**).

4.2. Influence of Microstructure on Hardness

Figure 12a illustrates the relationship between the mean area of primary austenite grains and material hardness, whereas Figure 12b presents the correlation between hardness and ferrite volume fraction. The data indicate a tendency for hardness to increase with an increase in the mean cross-section area of primary austenite grains and to decrease with an increase in the ferrite volume fraction.

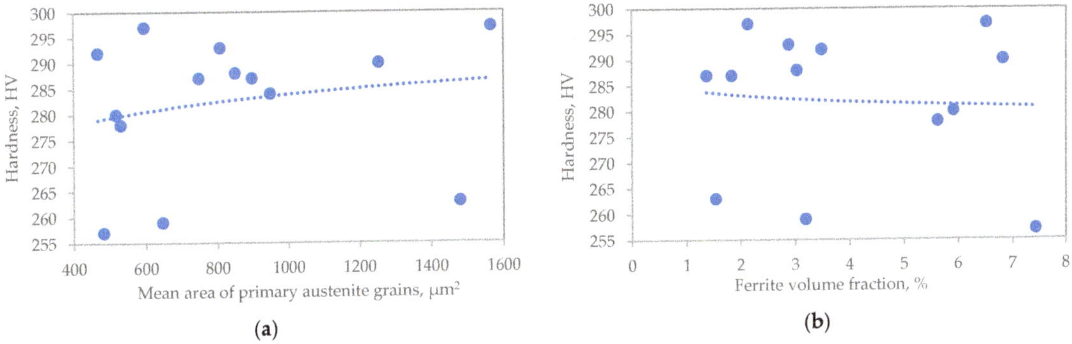

Figure 12. Dependence of hardness on: the cross-sectional area of primary austenite grain (**a**), ferrite volume fraction (**b**).

4.3. Analysis of the Correctness of the Assumptions Regarding the Occurrence and Nature of the Weld, the Heat-Affected Zone, and the Parent Material

Macroscopic observations of samples taken from the wheel flange no. 1–5 and 8–13 are characterized by the so-called area modified as a result of post-casting processing (Figure 8). Most likely, these are places where casting defects were repaired by welding. To select regenerative weld (RW), the obtained contents of manganese, silicon, and chromium of samples taken from the wheel tread edge area were compared with the measurement values of samples taken from the flange (Figure 13). Preliminary analysis showed that samples taken from the treat edge area from wheels 1–4, 8, 10, and 12 most likely contain only foundry weld (W). Differences in chemical composition can be seen for samples from wheels 5–7, 9, 11, 13, and 14, which may indicate that the welding was carried out only after the wheels (RW) were worn out. These assumptions were confirmed by microstructural documentation for samples from wheels 5, 9, 11, and 13 (Figure 10). Wheels 6 and 7 were rejected based on previous macroscopic observations that indicated no welding was used (Figure 8). In the microstructural documentation for the sample from wheel nr 14, a microstructure characteristic of the parent material was observed (Figure 10). The low hardness also confirmed that the welding process was not applied to wheel 14, and thus, the analysis will be focused on samples from wheels 5, 9, 11, and 13.

Figure 14 shows the welding zones along with the parent material after a comprehensive analysis, including chemical composition, microstructure, and hardness. Based on macro- and microstructural observations, each sample was also classified according to the degree of deformation (large, medium, small).

Figure 13. Contents of manganese (**a**), chromium (**b**), and silicon (**c**) in the modified area for samples taken from the flange and in the assumed pad weld for samples taken from the tread edge area.

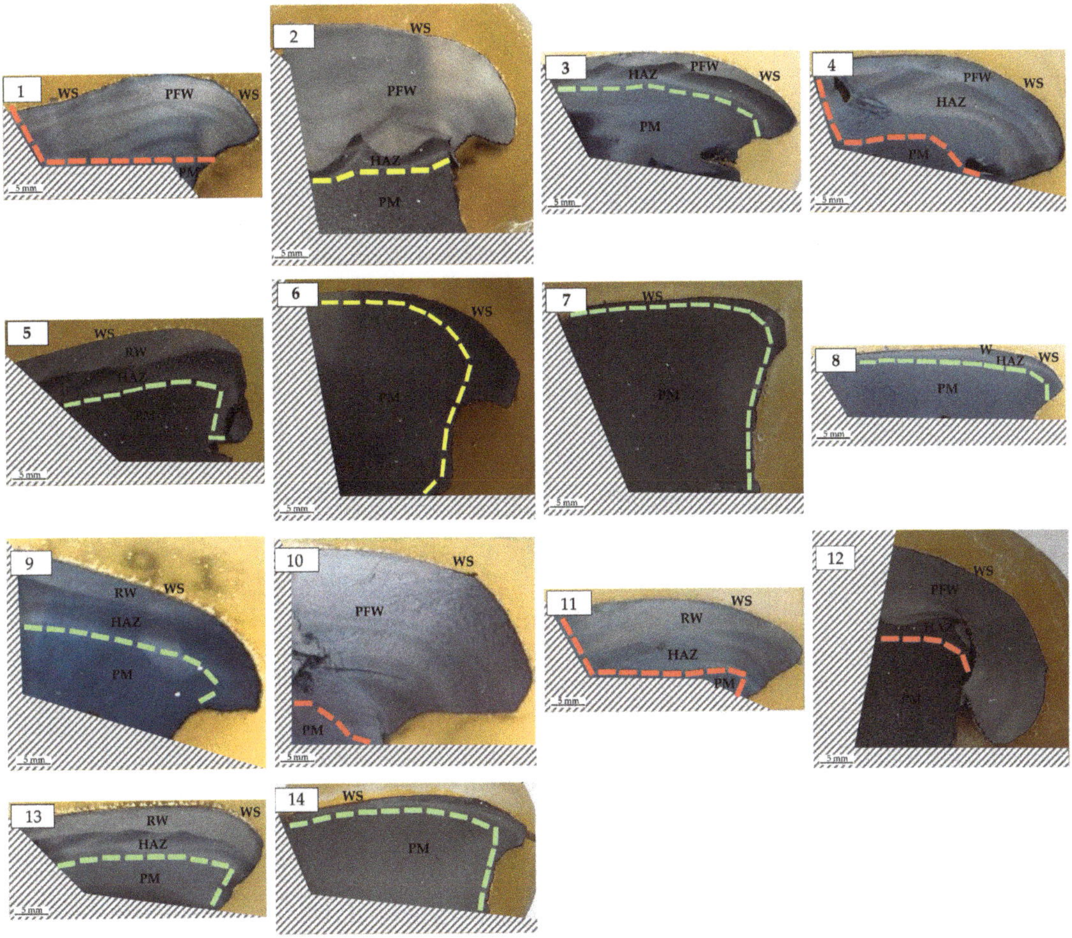

Figure 14. Macro image of etched metallographic sections with marked deformation degree (large deformation—red line, medium deformation—yellow line, small deformation—green line) and areas: repair weld (RW), post foundry weld (PFW), heat-affected zone (HAZ), parent material (PM), working surface (WS). (**1–14**): wheel number.

4.4. Influence of Chemical Composition on Weld Hardness

Figure 15 illustrates the impact of chemical composition on the hardness of the regenerative weld. There is a modest tendency for the hardness of the tested material to increase with increasing content of manganese, chromium, and silicon. This phenomenon can be attributed to the positive influence of Cr and Mn on the potential formation of banite, as well as the solid solution strengthening effect provided by silicon.

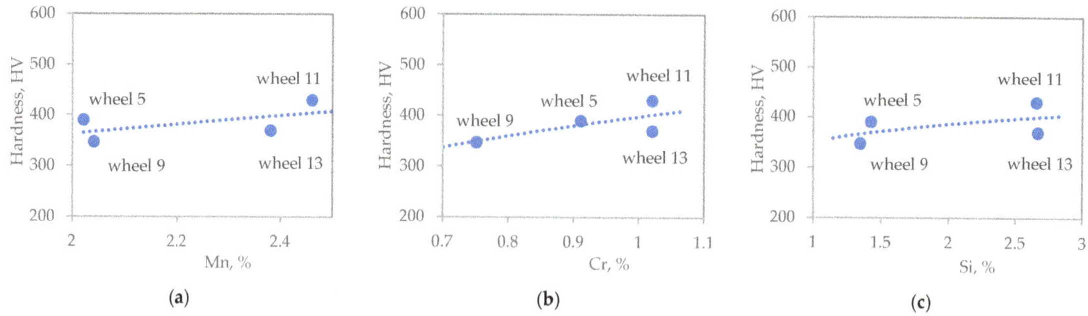

Figure 15. Effect of manganese (**a**), chromium (**b**), and silicon (**c**) on the weld hardness.

4.5. Relationship Between Chemical Composition and Deformation Degree

Figure 16 shows the dependence of the degree of deformation on the content of manganese, chromium, and silicon. It can be seen that elevated concentrations of these elements correspond to higher deformation, a phenomenon that is influenced by numerous factors. This is most likely attributed to the fact that manganese, chromium, and silicon reduce the material's susceptibility to abrasive wear and increase the tendency to wear through plastic flow.

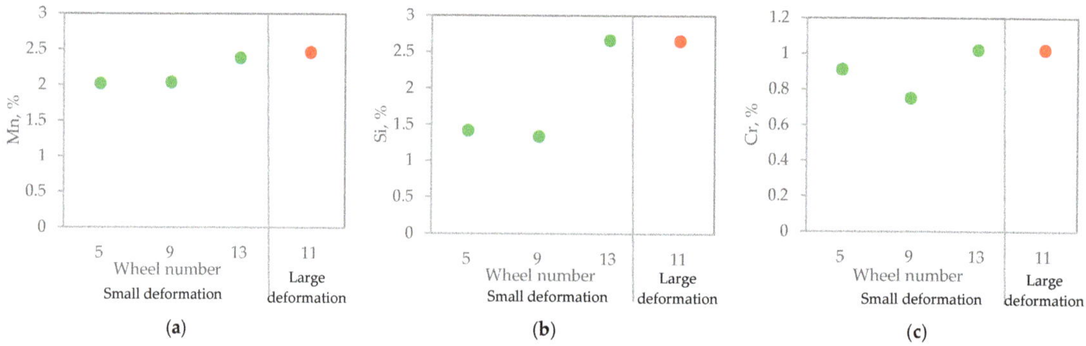

Figure 16. Effect of manganese (**a**), chromium (**b**), and silicon (**c**) on the repair weld deformation degree (large deformation—red dots, small deformation—green dots).

4.6. Influence of Deformation Degree on Hardness

The influence of the deformation degree of the regenerative weld on its hardness is shown in Figure 17. It can be seen that the hardness values correspond to a higher deformation degree, which can be related to the so-called deformation strengthening. The material strengthens due to the slip of dislocations on intersecting slip planes. The accumulation of dislocations at these specific locations leads to a decrease in ductility, accompanied by an increase in strength and hardness.

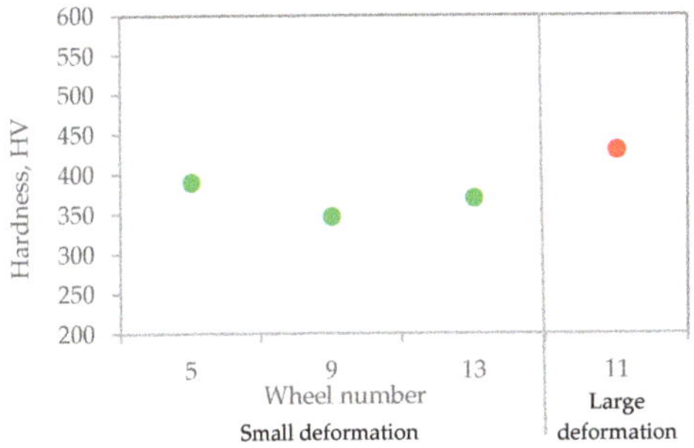

Figure 17. Effect of repair weld deformation degree on hardness. (large deformation—red dots, small deformation—green dots).

5. Conclusions

The findings of the research and analysis conducted allow for the following conclusions to be drawn:

1. It was discovered that padding welds were conducted at two distinct stages: initially, during the repair of bare wheel castings, and subsequently, following the completion of post-operation repairs.
2. The microstructure of the weld is predominantly bainitic.
3. The primary mechanism of wear observed in bare wheel components, both in the vicinity of welds and in the parent material, is the plastic flow of the material. It resulted in a change in the geometry of the wheel tread edge area, making further operation of the wheel impossible. Plastic flow occurred towards the free surface, and the presence of regenerative welds in the area of the wheel working surface may indicate earlier wear of the wheels due to the fatigue mechanism. Cracking was initiated on the surface or just below the surface in the area of the maximum shear stress (Hertz), and these cracks most likely resulted in defects that were filled by surfacing.
4. The hardness of the material is dependent upon the chemical composition, the microstructural components, and the degree of plastic deformation.
    - The addition of alloying elements results in increased hardness;
    - The formation of non-equilibrium structures (bainite) also results in increased hardness;
    - The observed increase in hardness is a consequence of strain hardening due to exploitation.
5. Increasing the alloyability as a result of the padding weld helps to achieve greater plasticity of the material.
6. The proposed methodology for analyzing the research results enables the drawing of conclusions regarding the origin of the padding weld (repair of castings and regeneration after operation). It allows us to distinguish the results of repair processes of wear effects in a highly loaded friction node with unsteady lubrication conditions (environmental influence) from repair processes applied to castings of large construction components.

Author Contributions: Conceptualization. J.K.; methodology. Ł.F.; investigation. Ł.F. and J.K.; writing—original draft preparation. P.M.; writing—review and editing Ł.F and P.M.; supervision J.K.; validation, Ł.F. and P.M. All authors have read and agreed to the published version of the manuscript.

Funding: The research was financed by the Ministry of Education and Science (AGH—research subsidy No. 16.16.110.663).

Data Availability Statement: Data are contained within the article.

Acknowledgments: The authors would like to thank Jagoda Kaptur and Elżbieta Dudek for their support with the research.

Conflicts of Interest: The authors declare no conflict of interest.

References

1. Szudyga, M. Diagnostics of Fatigue by Magnetic Methods of Steel Used in Railway Wheels and Rims of Railway Wheel Sets. Ph.D. Thesis, Silesian University of Technology, Katowice, Poland, 2011.
2. Zhou, H.; Zhang, H.; Yang, C. Hybrid-model-based intelligent optimization of ironmaking process. *IEEE Trans. Ind. Electron.* **2019**, *67*, 2469–2479. [CrossRef]
3. Adilson de Castro, J.; Medeiros, G.A.d.; Oliveira, E.M.d.; de Campos, M.F.; Nogami, H. The Mini Blast Furnace Process: An Efficient Reactor for Green Pig Iron Production Using Charcoal and Hydrogen-Rich Gas: A Study of Cases. *Metals* **2020**, *10*, 1501. [CrossRef]
4. Łędzki, A.; Zieliński, K.; Klimczyk, A. *Basics of Manufacturing and Processing Technology. Part V. Steelmaking*; AGH University of Kraków: Cracow, Poland, 2015.
5. *EN 13262:2020*; Railway Application. Wheelsets and Bogies. Wheels. Product Requirements. European Standard: Brussels, Belgium, 2020.
6. Available online: https://www.transportszynowy.pl/Kolej/zestawykolowe-kola (accessed on 30 November 2024).
7. Zani, N.; Mazzù, A.; Solazzi, L.; Petrogalli, C. Examining Wear Mechanisms in Railway Wheel Steels: Experimental Insights and Predictive Mapping. *Lubricants* **2024**, *12*, 93. [CrossRef]
8. Shi, Z.; Nencioni, L.; Meli, E.; Ding, H.; Wang, W.; Andrea, R. Effect of Material Hardness Ratio on Wear and Rolling Contact Fatigue: Development and Validation of New Laws. *Wear* **2023**, *514–515*, 204561. [CrossRef]
9. Zhou, G.Y.; Liu, J.H.; Wang, W.J.; Wen, G.; Liu, Q.Y. Study on the fatigue and wear characteristics of four wheel materials. *J. Mod. Transport.* **2013**, *21*, 182–193. [CrossRef]
10. Pointner, P. High strength rail steels—The importance of material properties in contact mechanics problems. *Wear* **2008**, *265*, 1373–1379. [CrossRef]
11. Mendez, P.F.; Barnes, N.; Bell, K.; Borlea, S.D.; Gajapathi, S.S.; Guest, S.D.; Izadi, H.A.; Gol, K.; Wood, G. Welding processes for wear resistant overlays. *J. Manuf. Process* **2014**, *16*, 4–25. [CrossRef]
12. SaifulAkmal, M.N.; Wahab, M.N. Characterization of UIC-54 Rail Head Surface Welded by Hardfacing Using Flux-Cored Steel Wire. In *Recent Trends in Manufacturing and Materials Towards Industry 4.0*; Lecture Notes in Mechanical Engineering; Osman Zahid, M.N., Abdul Sani, A.S., Mohamad Yasin, M.R., Ismail, Z., Che Lah, N.A., Mohd Turan, F., Eds.; Springer: Singapore, 2021. [CrossRef]
13. Goo, B.-C.; Seo, J.-W.; Lee, Y.-J. Effect of Welding Polarity on Mechanical Properties of Submerged Arc Welded Railway Vehicle Wheels. *Metals* **2022**, *12*, 1381. [CrossRef]
14. *ISO 1005-1:1994*; Railway Rolling Stock Material. Part 1: Rough-Rolled Tyres for Tractive and Trailing Stock—Technical Delivery Conditions. International Standard: Geneve, Switzerland, 1994.
15. *PN-K-88202:1996*; Railway Rolling Stock-Steel Castings-Requirements and Tests. Polish Standard: Warsaw, Poland, 1996.
16. Bogacz, R.; Świderski, Z. Simulation of exploitation conditions induced damage of rolling surface of rails and wheels. *Symulacja w Badaniach i Rozwoju* **2010**, *1*, 119–129.
17. Vakkalagadda, M.; Vineesh, K. Causes and failure forms of railway Wheels. *Mater. Today Proc.* **2024**, *98*, 97–101. [CrossRef]
18. Goo, B.C.; Hwang, S.; Choi, S.Y.; Lee, Y.J. Worn-Wheel Restoration by Welding and Evaluation of Mechanical Properties. *J. Korean Soc. Railw.* **2018**, *21*, 241–248. [CrossRef]
19. Kim, B.-R.; Moon, K.Y.; Cho, J.S.; Lee, Y.J. Applicability Verification of Restoration by Welding to Extend the Lifetime of Railway Wheel. *Acad. Soc. Appropr. Technol.* **2023**, *9*, 81–88. [CrossRef]
20. Maglio, M.; Vernersson, T.; Nielsen, J.C.O.; Ekberg, A.; Kabo, E. Influence of railway wheel tread damage on wheel–rail impact loads and the durability of wheelsets. *Rail. Eng. Sci.* **2024**, *32*, 20–35. [CrossRef]
21. The Welding Institute (TWI). Remanufacture of Rail Wheels: Aurora Project. Available online: https://www.twi-global.com/media-and-events/insights/aurora-project-remanufacture-of-rail-wheels (accessed on 11 January 2024).
22. Coo, B.C.; Lee, Y.J. Railway Vehicle Wheel Restoration by Submerged Arc Welding and Its Characterization. *Sci* **2020**, *2*, 33. [CrossRef]

23. Gajvoronsky, A.A.; Poznyakov, V.D.; Sarzhevsky, V.A.; Vasiliev, V.G.; Orlovsky, V.Y. Influence of thermo-deformational cycle of hardfacing on the structure and properties of railway wheels at their reconditioning. *Paton Weld. Sci. Tech.* **2010**, *5*, 15–18. [CrossRef]
24. Markisha, L.I.; Poznyakov, V.D.; Gajvoronsky, A.A.; Berdinkova, E.N.; Alekseenko, T.A. Structure and properties of railway wheel surface after restoration surfacing and service loading. *Paton Weld. J.* **2015**, *5–6*, 96–100. [CrossRef]
25. Mičian, M.; Winczek, J.; Gucwa, M.; Koňár, R.; Málek, M.; Postawa, P. Investigation of Welds and Heat Affected Zones in Weld Surfacing Steel Plates Taking into Account the Bead Sequence. *Materials* **2020**, *13*, 5666. [CrossRef]
26. Janiczak, R.; Pańcikiewicz, K. Laser welding of austenitic ferrofluid container for the KRAKsat satellite. *Weld. World* **2021**, *65*, 1347–1357. [CrossRef]
27. Pańcikiewicz, K.; Radomski, W. Lack of tightness analysis of concealed welded radiators. *Eng. Fail. Anal.* **2020**, *114*, 104579. [CrossRef]
28. Mi, G.-F.; Liu, Y.-L.; Zhang, B.; Fu, X.-Q.; Zhang, H.; Song, G.-X. Wear property of cast steel wheel material in rail truck. *J. Iron Steel Res. Int.* **2009**, *16*, 73–77. [CrossRef]
29. Getmanova, M.E.; Ilyukhin, D.S.; Nikulin, A.N.; Filippov, G.A. Composition and properties of steel in cast and forged railroad wheels. *Steel Transl.* **2017**, *47*, 70–77. [CrossRef]
30. *PN-H-84027-1:1994*; Railway Steel-Wheels-Types. Polish Standard: Warsaw, Poland, 1994.
31. *EN 10027-1:2016*; Designation Systems for Steels-Part 1: Steel Names. European Standard: Brussels, Belgium, 2016.
32. *EN 10027-2:2015*; Designation Systems for Steels-Part 2: Numerical System. European Standard: Brussels, Belgium, 2015.

lubricants

MDPI

Article

Modeling of the Dynamics of Conical Separate Plates in a Wet Multi-Disc Clutch

Qin Zhao [1], Biao Ma [1], Cenbo Xiong [1,*], Liang Yu [1], Bing Fu [2] and Shufa Yan [3]

1 School of Mechanical Engineering, Beijing Institute of Technology, Beijing 100081, China; yudewuhou@outlook.com (Q.Z.); mabiao@bit.edu.cn (B.M.); yuliang@bit.edu.cn (L.Y.)
2 School of Mechanical Engineering and Mechanics, Xiangtan University, Xiangtan 411105, China; fubing@xtu.edu.cn
3 Automotive Research Institute, China National Heavy Duty Truck Group, Jinan 250101, China; yanshufa1990@163.com
* Correspondence: xiongcenbo@bit.edu.cn

Abstract: Wet multi-disc clutches in transmission systems suffer from the incomplete separation of the friction components, which raises the drag torque and results in power loss and heightened fuel consumption. This incomplete separation arises from the force imbalance between resistance forces, such as the oil viscosity force, and the lack of an axial separating force. Therefore, providing an axial separating force is a potential solution to this problem. In this investigation, small-angle conical separate plates were designed which can provide the elastic restoring force during the separation process. Based on its structural properties, a model describing the clutch engagement and separation process was established. Through bench tests, the feasibility of the model was verified. The influence of the conical plate on the dynamics of the clutch was studied, including the influence of the separation gap, uniformity, and drag torque. Though the transmitted torque was reduced by 10.31% in the low-piston-pressure condition and by less than 2% in the high-piston-pressure condition, the problem of incomplete separation was successfully resolved. The results show that when applying the conical plates, the separation time was reduced by 18.78%, with a 25.31% increase in the uniformity of the gaps. Accordingly, the drag torque was reduced by 37.73%.

Keywords: wet multi-clutch; drag torque; failure prevention; numerical modeling

1. Introduction

Wet multi-disc clutches find widespread usage in various types of facilities, including in ocean-going ships, helicopters, heavy vehicles, and construction machinery, due to their compact structure, large transmission torque, and notable wear resistance [1,2]. In these clutches, power shifting and torque regulation are realized by the engagement and separation of friction components [3,4]. The working process of a wet clutch determines its stability and reliability, and thus is of great importance [5]. The engagement process takes place under high pressure and high speed difference conditions. During the separation process, the friction components are released when the piston pressure is reduced. Under the action of the oil film and rotating friction components, drag torque is generated. In ideal conditions, the friction components should move back to their original positions before engagement. However, in practice, the friction components may not completely separate, meaning that the drag torque is much higher than calculated. Accordingly, jamming and shift delay may occur [6].

The typical phenomenon of incomplete separation involves all the friction components sticking together after the separation process. As shown in Figure 1, though the piston can be pulled back by the return spring, there is no axial force to help the friction components move away from each other. Under the action of the oil viscosity force [7,8], the adsorption force caused by surface tension [9], and atmospheric pressure, the friction components stick together. Therefore, the narrowed gaps and oil film between friction pairs may increase the drag torque and power loss [10,11]. Robert [12] studied the influence of the pack clearance, oil flow and level, and clutch on the power loss from the drag torque. Lukas et al. [13] also used an experimental method to investigate the drag loss of a clutch. Leister et al. pointed out that the drag torque under separation status is a major source of energy loss [14]. Thomas et al. [15] warned that the significant loss in power that occurs under separation status should be given more serious consideration. Since the wet clutch is under separation status for most of the operating duration, the energy consumption and exhaust emissions generated by the drag torque under separation status make up a large proportion of all energy consumed. Furthermore, Hilpert et al. [16] indicated that if friction components cannot fully separate, rough contact may occasionally occur. Hou et al. [17] suggested that the mechanical contact caused by the oscillating motion was the reason for the sharp increase in torque.

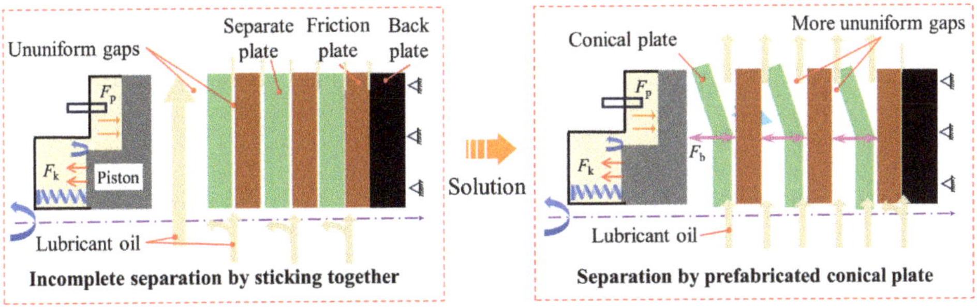

Figure 1. Schematic diagram of incomplete separation and the solution.

The solution to the incomplete separation is to provide an external force to forcibly separate the separate plate from the friction plate. Yu et al. [18] introduced and analyzed a waveform-separating spring, which can provide extra force to help pull apart the friction components, emphasizing the influence of the wavenumber on the load–deformation relationship. Erfanian-Naziftoosi et al. [19] verified that waveform-separating springs can retain their stiffness even after millions of load cycles. Li et al. [20] then analyzed multiple parameters of the separating spring, indicating that the radial deformation is related mostly to the wave height and width of the spring. Yan et al. [21] investigated the non-linear influence of the wave-shaped cushion plate and the deformed diaphragm spring on the engagement and disengagement characteristics in pull-type diaphragm spring clutches. Wang et al. [22] investigated the influence of the wavenumber and thickness of the separating spring on the engagement process. Unfortunately, the results showed that a pressure loss during engagement leads to a transfer torque loss, which was unexpected. Furthermore, additional components may cause the system to become more complex and undermine the compactness of the clutch. Wu et al. [23] studied a cone-shape friction plate and realized that the engaging time increased significantly. Zhao et al. [24] demonstrated that a conical buckled plate has a special property, in that it can provide a normal force in the axial direction. To utilize the elastic restoring force, the influence of the conical plate on the engagement and separation characteristics should be investigated.

Herein, in order to help separate the friction components and reduce the drag torque without adding extra elements, a conical plate was introduced. Based on the structural properties, a dynamic model featuring a conical plate rather than the traditional flat plate was established. The engagement and separation properties of the clutch system were analyzed via a numerical method and verified in an experiment. The influences of the cone angle on the separation time, gaps, and drag torque were studied, and the optimal layout of the conical plate is presented.

2. Numerical Model

2.1. Conical Separate Plate Design

Inspired by the buckling deformation of the separate plate [24], a conical separate plate was designed, as shown in Figure 2. The plate was circumferential uniform with deformation in the radial direction, presenting as a conical (saucer) shape. The initial deformation was w_0. The force was applied along the z direction, as shown by the red arrows in Figure 2. The structural parameters were as follows:

$$\begin{cases} \alpha = \arcsin \frac{w_0}{R_0 - R_i} \\ d = 2R_\mathrm{i} \\ D = d + 2(R_\mathrm{o} - R_\mathrm{i}) \cos \alpha \end{cases} \tag{1}$$

where α is the cone angle, indicating the conical degree; d and D are the inner and outer diameter of the conical plate, respectively; and R_i = 0.086 m and R_o = 0.125 m are the inner and outer radius of the flat plate, respectively. The number and size of male splines were the same as those of the flat plate. The material of the plate was 30CrMnSi stainless steel with a high elastic modulus of 216 GPa and a Poisson's ratio of 0.3. Its density was 7750 kg/m^3.

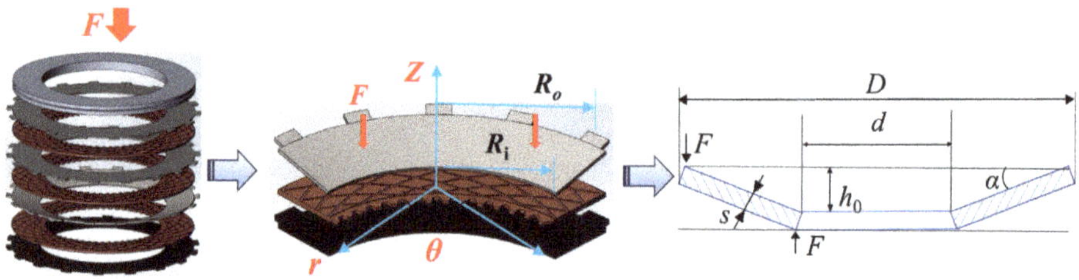

Figure 2. Conical separate plate.

The most notable difference between the flat separate plate and the conical separate plate was that the conical plate could provide axial force F_b (elastic restoring force) [24]:

$$\begin{cases} F_b = \frac{4\pi E s^4}{(1-v^2)D^2} a_1 \frac{w}{s} [(\frac{w_0}{s} - \frac{w}{s})(\frac{w_0}{s} - \frac{1}{2}\frac{w}{s}) a_0 + 1] \\ K_F = \frac{dF}{dw} = a_2 [3w^2 - 6w_0 w + 2w_0^2 + \frac{2s^2}{a_0}] \end{cases} \tag{2}$$

$$\begin{cases} a_0 = \frac{6(1-v^2)}{(c-1)^2 lnc} [\frac{1}{4}(c^2-1) - \frac{c^2}{c^2-1}(lnc)^2] \\ a_1 = \frac{1}{6}(\frac{c}{c-1})^2 lnc \\ a_2 = \frac{2\pi E a_1 a_0 s}{(1-v^2)D^2} \\ c = D/d \end{cases} \tag{3}$$

where K_F is the stiffness of the conical plate; E is the elastic modulus; s = 0.002 m is the thickness; v is the Poison's ratio; and w and w_0 are the deflection and initial deflection, respectively.

Since stiffness is a quadratic function of deflection w, the discriminant deciding the monotonicity of the plate was obtained as:

$$\Delta_1 = 36w_0^2 - 24\left(w_0^2 + \frac{s^2}{a_0}\right) = 12(w_0^2 - \frac{2s^2}{a_0}) \tag{4}$$

From Equation (4), if the condition $w_0/s < (2/a_0)^{0.5}$ is satisfied, the stiffness is positive during the deformation process. Otherwise, there will be negative stiffness. Because the conical plate should provide enough force during the whole separation process, the force should be monotone increasing. Thus, the cone angle should satisfy the moderate size requirement.

2.2. Dynamic Model

The structure of a wet multi-disc clutch is illustrated in Figure 3. For a N pair wet clutch, there were $N/2$ separate plates, $N/2$ friction plates, one back plate, and one piston involved in the movement. Using two-dimensional coordinates, with the origin located on the left side of the back plate, the location of each component was 0 to N. The subscripts p, s, and f denote the piston, separate plate, and friction plate, respectively. The force governing equations along the axial direction is:

$$\begin{cases} F_{v0} + F_{c0} + F_{k0} + F_{R0} - F_{b0} - F_p - F_{d0} - F_{imp} - F_{cen} = m_p\ddot{x}_1 \\ F_{vi} + F_{ci} + F_{bi} + F_{R(i-1)} - F_{v(i-1)} - F_{c(i-1)} - F_{b(i-1)} - F_{di} - F_{si} - F_{Ri} = m_i\ddot{x}_i \\ \vdots \\ F_{vN} + F_{cN} + F_{RN} - F_{v(N-1)} - F_{c(N)} - F_{b(N-1)} - F_{sN} = m_N\ddot{x}_N \end{cases} \tag{5}$$

where x, its derivative, and its second derivative denote the displacement, speed, and acceleration; m is the mass; subscripts p, s, and f denote the piston, separate plate, and friction plate, respectively; F_k is the force of the return springs; F_R is the resistance force which prevents friction components from fully separation; F_b is the elastic restoring force; F_p is the hydraulic force provided by the piston, which tightly compresses the friction components; F_d is the damping force; F_{imp} is the impact contact force; F_{cen} is the centrifugal oil force, mainly acting when separation occurs; F_c is the asperity rough contact force; F_v is the viscus force; and F_s is the spline friction force. They could be calculated as:

$$\begin{cases} F_{k0} = k_{spr} \cdot (l + X_0 - x_0) \\ F_p = p \cdot \pi({R_o}^2 - {R_i}^2) \\ F_{d0} = c_0 \cdot \dot{x}_0 \\ F_{cen} = 0.25 \cdot \rho \cdot \omega^2 \cdot [{R_4}^4 - {R_1}^4 - 2 \cdot {R_0}^2 \cdot ({R_4}^2 - {R_1}^2)] \\ F_{imp} = K_0 \cdot \xi^n [1 + \frac{3(1-e^2)}{4} \cdot \frac{\dot{\xi}}{\dot{\xi}_0}] \\ F_s = \mathrm{sign}(\dot{x}) \frac{\mu_s M_s}{R_s \cos \alpha_s} \end{cases} \tag{6}$$

where k_{spr} is the stiffness of the return spring, l is the initial compression length of the return spring, and x_0 and X_0 are the axial and maximum axial displacement of the piston, respectively; ξ is the impact compression and equals $X_0 - x_0$; c_0 is the damping between the piston and the cylinder liner; ρ and ω are the density and rotational speed of the lubrication oil, respectively, R_0 to R_4 are the geometric parameters of the piston (Figure 3b); K_0 is the impact stiffness; and R_s, α_s, and μ_s are the pitch circle radius, the pressure angle of the spline, and the spline friction coefficient, respectively.

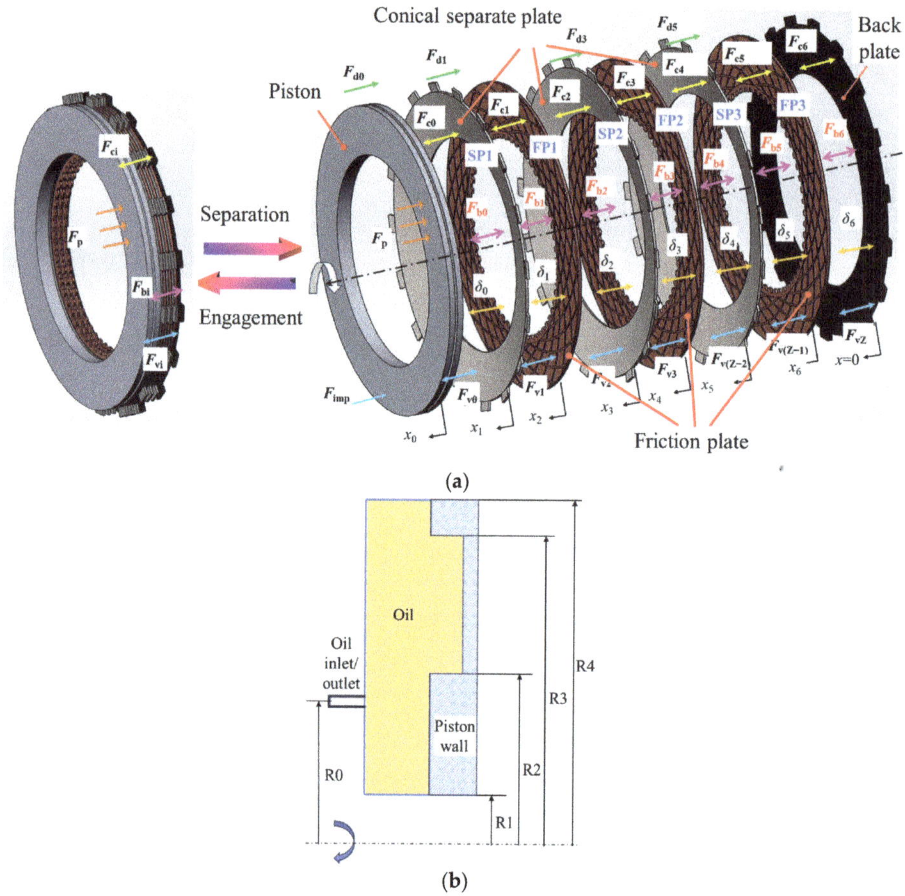

Figure 3. A schematic of the clutch system. (**a**) The dynamics of the clutch; (**b**) the structure of the piston.

After pressure relief, the elastic restoring force pushed the adjacent components away. However, resistance force tried to make the friction components stay together:

$$\begin{cases} F_R = F_d + F_{cap} \\ F_d = \eta \cdot v/h \cdot \pi(R_o^2 - R_i^2) \\ F_{cap} = 2 \cdot \pi \cdot R_{lb} \cdot \gamma \cdot \cos\theta \end{cases} \tag{7}$$

where F_d is the disengagement resistance force calculated by the thin-layer fluid separation model [13]; F_{cap} is the adsorption force caused by surface tension [25]; η is the viscosity of the oil; v is the separating velocity; h is the thickness of the oil film; R_{lb} is the radius of the liquid bridge between the friction and separate plate; γ is the surface tension coefficient of oil; and θ is the contact angle between the oil and the solid surface.

In response to the displacement change, the gaps between each pair were as follows:

$$\begin{cases} \delta_i = x_i - x_{(i+1)} - s_i \\ \delta_N = x_N \end{cases} \tag{8}$$

where s_i is the thickness of the separate plate or the friction plate.

2.3. Contact and Lubrication Model

As is shown in Figure 4, the actual contact area A_c only took up a small proportion of the nominal contact area A_n of the surface and can be expressed as follows [26]:

$$A_c = \kappa\pi^2(N\beta\sigma)^2 A_n F_2(H) \tag{9}$$

where F_2(H) is the distribution function of asperities.

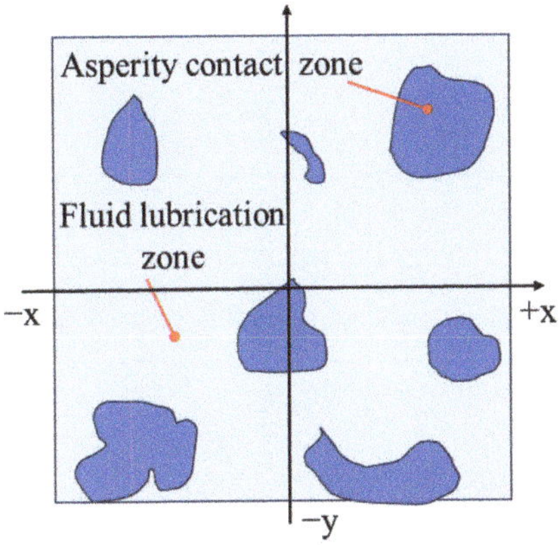

Figure 4. A sketch of the contact model.

In general, part of the asperities experienced plastic deformation, while other parts experienced elastic deformation. Index κ can represent the degree and $1 \le \kappa \le 2$. K could be used to calculate the elastic–plastic deformation status:

$$\begin{cases} K = \frac{E'}{H_B}\sqrt{\frac{\sigma}{\beta}} \\ \frac{1}{E'} = \frac{1}{2}\left(\frac{1-v_1^2}{E_1} + \frac{1-v_2^2}{E_2}\right) \end{cases} \tag{10}$$

where E' is the equivalent elastic modulus of the two contact surfaces, while E_1 and E_2 denote the elastic modulus of the two surfaces, respectively; v_1 and v_2 are the Poisson's ratio of the two surfaces, respectively; H_B is the Brinell hardness; and β, σ, and N are the radius, height, and the density of asperities, respectively.

The distribution of the asperities can be regarded as the Gaussian surface assumption [27]:

$$\varphi(z) = \frac{1}{\sqrt{2\pi}} e^{(-\frac{z^2}{2})} \tag{11}$$

Thus:

$$F_n(H) = \int_H^\infty (z - H)^n \varphi(z) dz \tag{12}$$

The contact pressure of the asperities can be calculated by the GT model [28]:

$$p_c = \begin{cases} K'E' \cdot 4.4086 \cdot 10^{-5} \cdot (4 - H)^{6.804}, \ H < 4 \\ 0 \qquad\qquad , H \ge 4 \end{cases} \tag{13}$$

where H is the film-thickness ratio $H = h/\sigma$; h is the film thickness; and the contact index K' can be calculated as:

$$K' = \frac{8\sqrt{2}}{15}\pi(N\sigma\beta)^2\left(\frac{\sigma}{\beta}\right)^{1/2} \tag{14}$$

The contact ratio C can be expressed as:

$$C = \kappa\pi^2(N\sigma\beta)^2[0.5 \cdot (H^2+1) \cdot \mathrm{erfc}(H/\sqrt{2}) - H/\sqrt{2\pi}e^{-\frac{H^2}{2}}] \tag{15}$$

The rough contact force is:

$$F_c = \pi(R_o^2 - R_i^2)p_c A_g C \tag{16}$$

where A_g is the ratio of the non-grooved area.

The Reynolds equation, considering the inertial force of the oil film, was established as follows [29]:

$$\begin{cases} \frac{\partial p}{\partial r} = \frac{\partial}{\partial z}(\eta\frac{\partial u_r}{\partial z}) + \rho r\omega^2 \\ \frac{\partial p}{r\partial\theta} = \frac{\partial}{\partial z}(\eta\frac{\partial u_\theta}{\partial z}) \\ \frac{\partial p}{\partial z} = 0 \end{cases} \tag{17}$$

where p is the oil-film pressure and ω is the rotational speed of the oil, which could be calculated as follows based on the laminar flow assumption due to the small Reynolds number:

$$\omega = \omega_f + (\omega_s - \omega_f) \cdot z/h \tag{18}$$

The boundary conditions were as follows:

$$\begin{cases} z = 0, u_r = 0, u_\theta = r\omega_s \\ z = h, u_r = 0, u_\theta = r\omega_f \end{cases} \tag{19}$$

Substituting, the velocity components could be calculated as

$$\begin{cases} u_r = \frac{(z^2-zh)}{2\eta} \cdot \frac{\partial p}{\partial r} + \frac{\rho r}{\eta}\Big[\frac{1}{2}\omega_s^2(hz - z^2) + \frac{(\omega_f-\omega_s)}{3h}\omega_s(h^2z - z^3) \\ \qquad + \frac{(\omega_f-\omega_s)^2}{12h^2}(h^3z - z^4)\Big] \\ u_\theta = 2\eta r \cdot \frac{\partial p}{\partial\theta}(z^2 - zh) + \frac{rz}{h}(\omega_f - \omega_s) + r\omega_s \end{cases} \tag{20}$$

The average oil film pressure between the friction pairs was

$$\begin{cases} \overline{p} = \frac{B}{4A}(r^2 - R_o^2) + \frac{3\eta}{A}\frac{\partial\overline{h}}{\partial t}(r^2 - R_o^2) + \ln\frac{r}{R_o}\left(\frac{B}{4A} + \frac{3\eta}{A}\frac{\partial\overline{h}}{\partial t}\right)\frac{R_o^2-R_i^2}{\ln R_i - \ln R_o} \\ A = \phi_r h^3 + 12\psi d_m \\ B = \frac{\phi_r\rho h^3}{5}(3\omega_f^2 + 4\omega_s\omega_f + 3\omega_s^2) \end{cases} \tag{21}$$

where ϕ_r is the pressure flow factor:

$$\phi_r = 1 - 0.9 \cdot e^{-0.56H} \tag{22}$$

The average oil-film pressure was integrated into the fluid lubrication area, and the viscus force between the friction pairs was deduced as:

$$F_v = \iint_{A_{rel}} \overline{p} = \pi\left\{\frac{B}{2A} + \frac{3\eta}{A}\frac{dh}{dt}[1 + \mathrm{erf}(\frac{h}{\sqrt{2}\sigma})]\right\} \cdot \left[\frac{R_i^4 - R_o^4}{4} - \frac{(R_o^2 - R_i^2)^2}{4(\ln R_i - \ln R_0)}\right] \cdot (1 - A_{rel} \cdot C) \tag{23}$$

2.4. Torque Model

According to Newton's law of internal friction, the viscous torque is expressed as [30]:

$$M_v = \int_0^{2\pi}\int_{R_i}^{R_o}\eta\frac{r^2\Delta\omega}{h}\mathrm{rdrd}\theta \tag{24}$$

Further, it could be calculated as:

$$M_v = \frac{\pi}{2} \cdot (1 - A_{rel}C) \int_0^{2\pi} \int_{R_i}^{R_o} \eta \cdot (\phi_f + \phi_s) \cdot \frac{r^2 \Delta\omega}{h} r dr d\theta \tag{25}$$

where ϕ_f and ϕ_s are the shear stress coefficients.

The rough contact torque is expressed as [31]:

$$M_c = A_c \cdot C \cdot \mu \cdot \int_0^{2\pi} \int_{R_i}^{R_o} p_c r^2 dr d\theta \tag{26}$$

The total friction torque is the sum of the viscous and rough contact torque:

$$M_f = M_v + M_c \tag{27}$$

3. Experiments

3.1. Apparatus

To verify the feasibility of the numerical method and obtain more detailed torque information, the long-time-sliding experiment was arranged. As is shown in Figure 5, the test bench was composed of a control system, a data acquisition system, a clutch pack, a torque sensor, etc. The clutch pack contained the piston, back plate, friction plates, and separate plates. In the six-pair system, N = 6, meaning that there were three separate plates and three friction plates. The torque transducer could record the torque data during the sliding and separation.

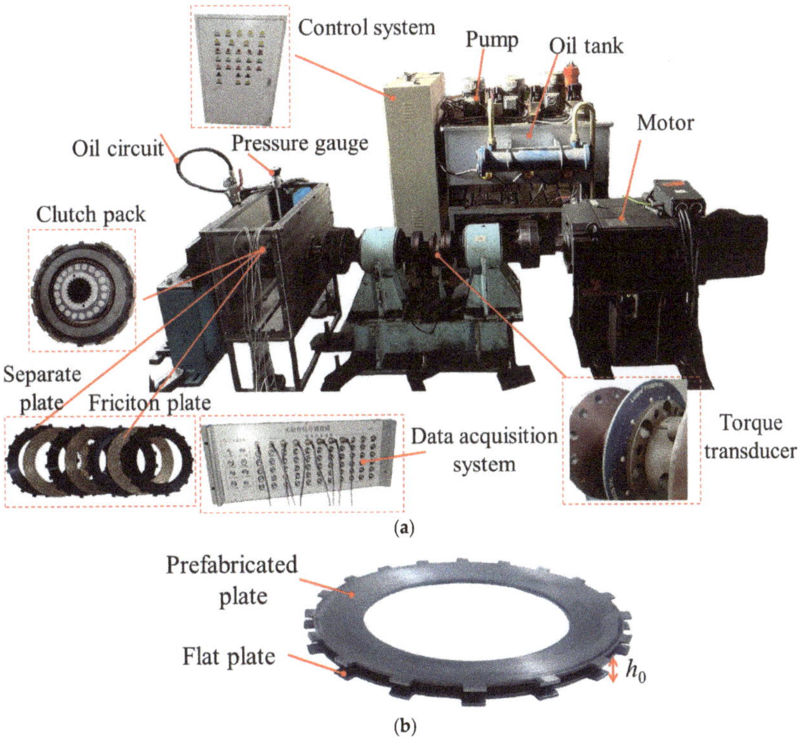

Figure 5. Experimental apparatus. (**a**) Test bench; (**b**) conical plate.

To guarantee the compactness of the clutch, the total reserved clearance of the clutch was 3.5 mm. In every case, all three separate plates were replaced with the conical plate, except for the original design. Since the total deformation of all the plates had to be smaller than the reserved clearance, a cone angle of 0.5°, 1°, or 1.5° was chosen to represent slight, moderate, or large conical degrees, respectively, with corresponding total initial deformations of 1.02 mm, 2.04 mm, and 3.06 mm, respectively. The arrangement of the friction components is shown in Figure 6. The cases are referred to as 0°, 0.5°, 1°, and 1.5°, respectively.

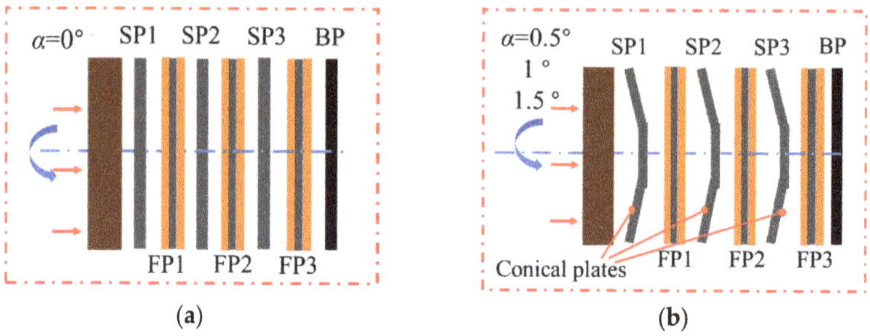

Figure 6. Experiment and simulation conditions. (**a**) C-0; (**b**) C-1 to C-3.

3.2. Methods

The simulation and experiment conditions are shown in Table 1 and Figure 6. For each cycle, the total time was 60 s. The piston pressure increased linearly at the beginning until reaching t = 0.5 s and was then maintained. Then, the piston pressure was revoked from t = 15 s. Controlled by the piston pressure, the working process of the clutch can be divided into three main stages: the engagement and stable sliding process (E), the separation process (S), and the separation status (R). Under the separation status, the piston pressure was reduced to zero; the rotational speed of the friction plates remained unchanged and was equal to the rotational speed of the motor.

Table 1. Conditions of the numerical method.

Case	Cone Angle (°)	Piston Pressure (MPa)	Rotational Speed (r/min)
C-0	0	0.25, 0.3, 0.35	300, 400, 500
C-1	0.5		
C-2	1		
C-3	1.5		

A numerical simulation was carried out using Matlab/Simulink 2022b, and the modules were programmed based on the models established in Sections 2.1–2.4. Based on the characteristics of the clutch engagement–separation process and the coupling of multiple variables, the ode23s solver was used in the simulation, the relative-tolerance error limit of which is smaller than 10^{-3}.

To verify the feasibility of the established model, the friction torque during sliding (abbreviated as the sliding torque) and under the separation status (drag torque) were obtained from the torque sensor and compared with the simulation results. To further verify the model, the performances under different values of piston pressure and rotational speed were simulated and tested. The influence of the cone angle was mainly analyzed under the conditions of 0.3 MPa and 400 r/min.

4. Results and Discussion

4.1. Model Verification

Figure 7 shows the comparison results of the friction torque between tests and simulations under 0.3 MPa 400 r/min. The red line indicates the experiment result and the black line indicates the simulation result. Figure 7 indicates that the larger the angle, the lower the friction torque during sliding (sliding torque). After the sharp increase and decrease caused by the overshoot of the motor, the sliding torque became basically stable in each case. The average sliding torque of the original design was 172.74 Nm, while the sliding torques for 0.5°, 1°, and 1.5° were 170.41 Nm, 160.34 Nm, and 155.22 Nm, respectively. The decrease came from the attenuative effect of the counterforce provided by the conical plate, which was comparable to when the piston pressure was as low as 0.3 MPa. The maximum difference between the simulation and experiments was 1.64%.

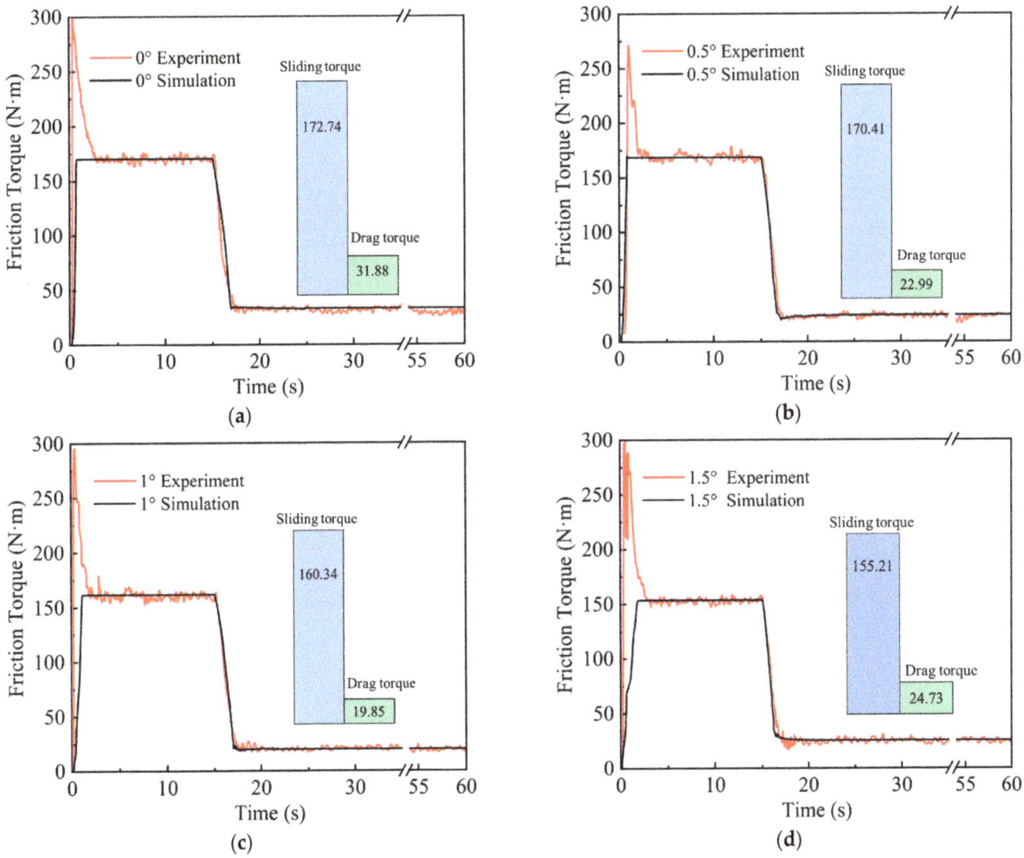

Figure 7. The torque results. (**a**) 0°; (**b**) 0.5°; (**c**) 1°; (**d**) 1.5°.

Figure 7 demonstrates that the larger the angle, the higher the drag torque. The average drag torque of the 0° case was 31.88 Nm, since the friction components stuck together even if the piston returned to its original position. The drag torques for 0.5°, 1°, and 1.5° were 22.99 Nm, 19.85 Nm, and 24.73 Nm, respectively. Because the oil films between pairs were rendered ineffective with the expansion of the conical plate, the drag torque was reduced. The maximum difference between the simulation and experiments was 4.50%.

To further verify the feasibility of the established model, the difference between simulations and experiments was obtained under different values of piston pressure and rotational speed (Figure 8). The maximum relative error of the sliding torque between experiments and simulations was 1.68%, while that of the drag torque was 6.76%, present at 0.35 MPa 400 r/min. Because of the force of the return springs, when an increase of 0.05 MPa from 0.25 MPa to 0.3 MPa and from 0.3 MPa to 0.35 MPa took place, the increased sliding torque was different. The drag torque had little influence over the piston pressure, while it was obviously affected by the rotational speed. The reason for this is that, when separated, the piston pressure was reduced to zero while the rotational speed of friction components was kept constant; the higher the speed, the larger the viscous force between each pair and the larger the drag torque. However, the sliding torque was minimally affected, mostly due to the difference in the friction coefficient.

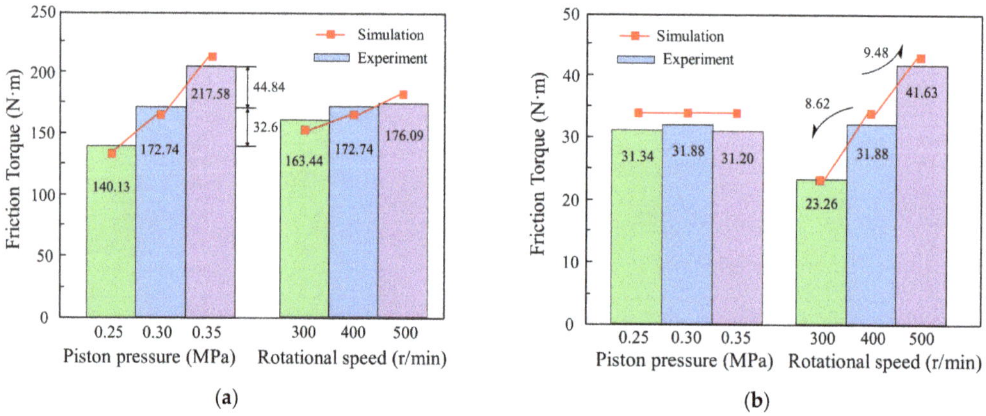

Figure 8. Further verification of the established model. (**a**) Sliding torque; (**b**) drag torque.

Since the relative error between the simulation and experiment was small and the trend of the simulation was close to that of the experiment, the viability of the numerical model was verified by the experiments.

4.2. Gap Characteristics

Based on the established model, the dynamic characteristics of the conical separate plate are discussed. The most notable differences when applying the conical plate were the width and uniformity of the gaps, which affected the lubrication and drag torque. Therefore, the difference in the gaps was studied. In addition, the engagement and separation time is discussed. The gaps between the plates underwent a cycle of narrowing–eliminating–expanding during the whole process. During the engagement process, the friction components moved axially and the speed difference narrowed to zero. During the separation process, the friction components separated and the gaps increased.

To quantitative analyze the influence on the dynamic characteristics, the coefficient of gap nonuniformity δ_u was defined as:

$$\delta_u = \frac{1}{6}\left(\sum_{i=0}^{6} \frac{1}{1+\delta_i/\delta^*}\right) \tag{28}$$

where δ^* = 0.5 mm is the ideal gap. A larger δ_u indicates higher nonuniformity.

The influence of the cone angle on the engagement and separation is obvious. As the angle increased from 0° to 1.5°, the engagement time increased from 0.79 s to 1.87 s. It was 137% longer for the 1.5° case (Table 2). The piston pressure needed to firstly overcome the elastic restoring force of the conical plates and then narrow the clearance. The restoring force is comparable with the piston force under low-pressure conditions and accordingly, the piston was slowed down. This is consistent with Wu's [23] results, in which a conical friction plate was found to delay the engagement.

Table 2. Time needed for engagement and separation.

	0°	**0.5°**	**1°**	**1.5°**
t_e (s)	0.79	0.90	1.18	1.87
t_s (s)	2.29	2.03	1.86	2.09
δ_0 (mm)	3.153	1.466	1.434	1.058
δ_u	0.9208	0.7315	0.6877	0.6528

Figure 9 illustrates the gaps during the engagement process. The gaps experienced a sharp drop when the applied force exceeded the force of the return springs. Meanwhile, the contact ratio of each pair grew exponentially. When the resultant force was stable, the gaps were nearly unchanged. In the original design $\alpha = 0°$, after $t = 0.79$ s, the gaps of all pairs were narrowed in order from S0 to S6 to an average of 10.55 μm. The average gap was narrowed from 9.65 μm to 9.18 μm with the increasing angle. This difference results from the structure of the conical plate. Although in macrography the plates were flattened when sliding, deformation occurred [24]. Therefore, deformation squeezed and narrowed the oil films between the contact pairs. The lubrication status was boundary lubrication due to the thin oil film.

Figure 10 demonstrates the gaps during separation. The position of each plate is shown in Figure 10 as SP1, SP2, and SP3. The gaps are also illustrated. Since the applied pressure was revoked, the gaps gradually reverted to their initial positions. The time needed for separation decreased from 2.29 s to 1.86 s, i.e., 18.78% shorter on account of the external elastic restoring force. The reason for the minimal difference was that the elastic restoring force was small compared with the piston force, so the length of the separation time was mostly due to the withdrawal of the piston. The elastic restoring force was much larger than the disengagement resistance force and the adsorption force; thus, the larger the angle, the larger the elastic restoring force, and the shorter the time needed for separation. However, when the angle exceeded 1°, the separation time was extended, because recovering from deformation also required time.

Unlike the engagement process, the time needed for separation varied from pair to pair. When $\alpha = 0°$, though, the separation times were close, because the time needed was fully dictated by the action of the piston, while that of the conical plates was decided by both the action of the piston and the conical plates themselves. Taking $\alpha = 1°$as an example, the times needed to reach stable gaps were t = 16.72 s, 17.07 s, 16.86 s, 17.08 s, 16.88 s, 16.88 s and 16.54 s. It was clear that the restoring side of the plate needed more time to become stable.

When there was no conical plate, δ_0 occupied most of the clearance, while δ_1 to δ_6 decreased in turn, with a smaller clearance (Figure 10a). Since the return spring was connected to the piston directly, when the piston pressure was revoked, the piston was pulled back to the original position. However, there was no external force impacting the rest of plates, so they could only separate through the centrifugal effect and could not move further. The lubrication statuses were mixture lubrication for all pairs except S0.

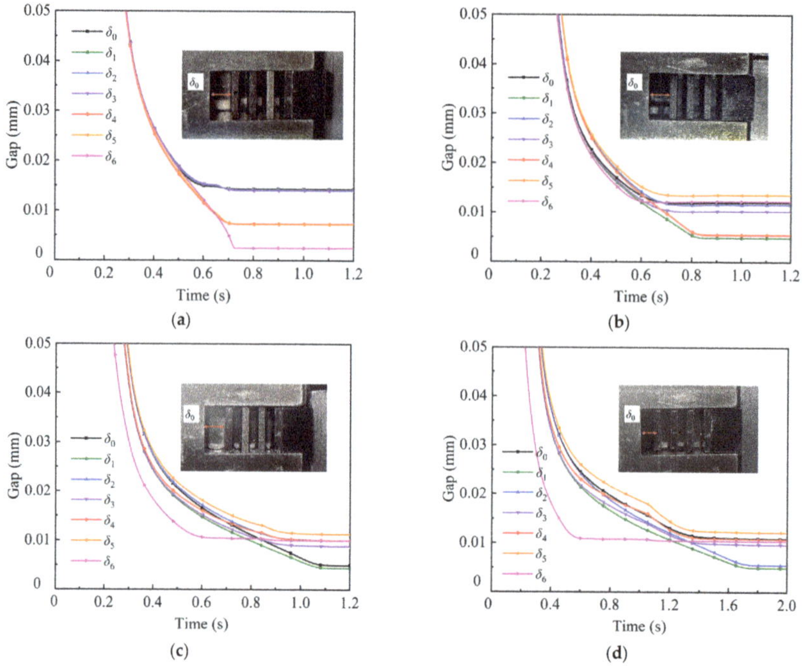

Figure 9. The influence of the conical plate on the engagement. (**a**) $\alpha = 0°$; (**b**) $\alpha = 0.5°$; (**c**) $\alpha = 1°$; (**d**) $\alpha = 1.5°$.

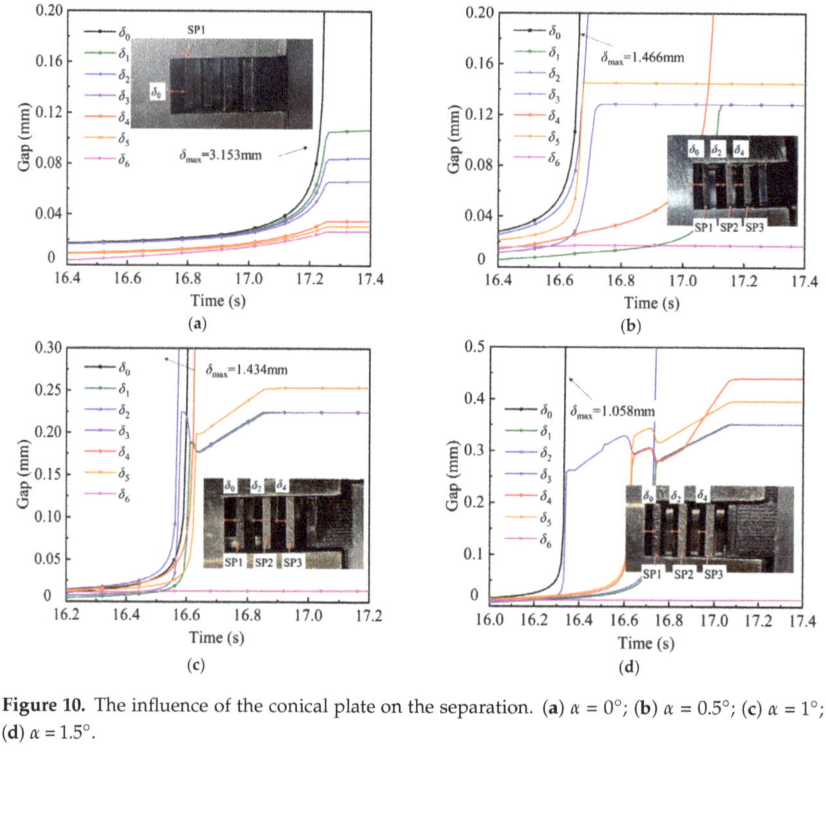

Figure 10. The influence of the conical plate on the separation. (**a**) $\alpha = 0°$; (**b**) $\alpha = 0.5°$; (**c**) $\alpha = 1°$; (**d**) $\alpha = 1.5°$.

When introducing the elastic restoring force, δ_0, δ_2, δ_4 became the largest. For example, the gaps for $\alpha = 1°$ were 1.434 mm, 0.225 mm, 0.701 mm, 0.224 mm, 0.647 mm, 0.254 mm, and 0.014 mm (Figure 10c). The elastic restoring force provided by the conical separate plate forced the separate plates and friction plates to move axially rather than only relying on the centrifugal force. However, the gap δ_6 was still as small as in the original design due to the lack of the elastic restoring force for FP3 or the back plate. The nonuniformity of gaps also decreased with the increasing angle until $\alpha = 1.5°$. The more uniform gaps enabled the lubricating oil to be allocated to each pair. The lubrication status was hydrodynamic lubrication for all pairs except S6, since gap δ_6 was too small.

4.3. Torque Characteristics

Figure 11 demonstrates the influence of conical plates on the friction torque during engagement and sliding (E). The influence of the cone angle on the sliding torque and drag torque is illustrated. After the piston chamber had been filled with oil, the friction torque increased quickly during the E1 stage. At stage E2, the viscous torque decreased to nearly 0, while the rough contact torque grew to its maximum, which can be referred to as the sliding torque. Since t_e, the sliding between separate plate and friction plates became stable. Figure 11 illustrates that the larger the cone angle, the smaller the sliding torque. The difference arose from the elastic restoring force provided by the conical plate. When the piston pressure was as small as 0.3 MPa, the counterforce between cases was not negligible. Specifically, the difference between the 0° and 1.5° cases was about 17.53 N·m, 10.31%. When the piston pressure was high, the difference was negligible. For example, when $p = 1.2$ MPa, the difference between the 0° and 1.5° cases was less than 2%. A high piston pressure is extensively employed in heavy-duty vehicles and facilities, and the clutch continues to increase the piston pressure and rotational speed to increase the energy density. Therefore, the advantages of the conical plate can outweigh the disadvantages, especially in high-piston-pressure conditions now and in the future.

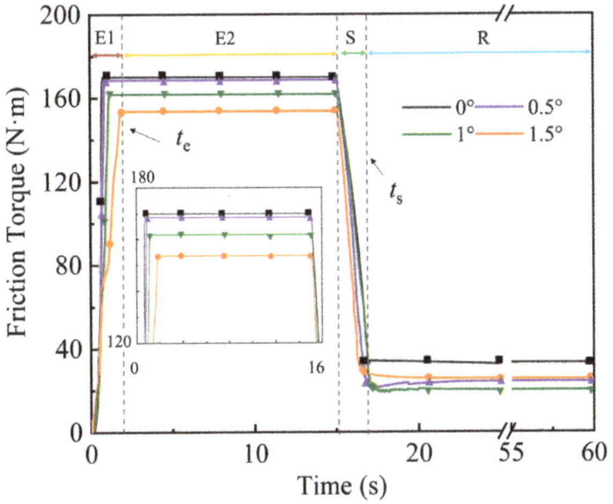

Figure 11. The friction torque during the cycle.

Then, during the separation process (S), the rough contact torque quickly declined while the viscous torque took effect. Under the separation status (R), the viscous torque increased a little and became stable; this can be referred to as the drag torque. Figure 12b demonstrates the influence of the cone angle on the drag torque. In the original design

(α = 0°), the friction components were stuck together, so that the drag torque was the highest among cases.

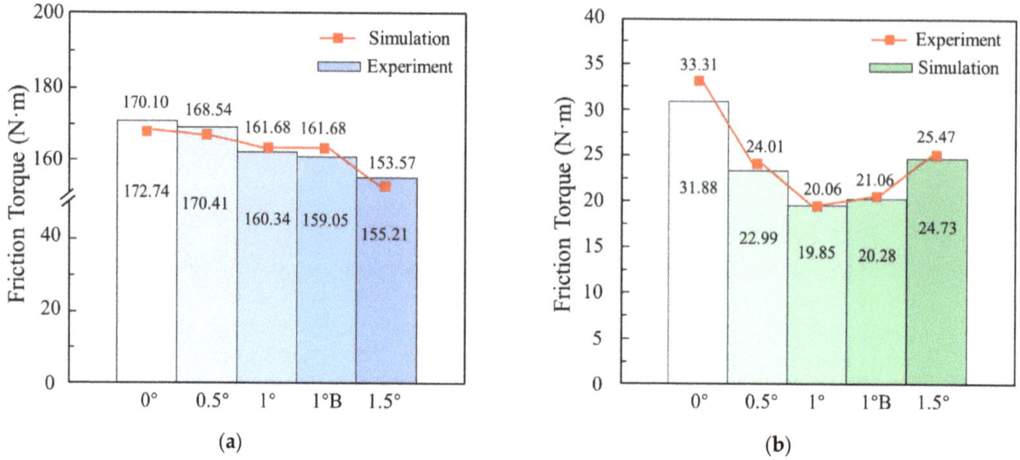

Figure 12. The influence of the conical plate on the sliding torque and drag torque. (**a**) Sliding torque; (**b**) drag torque.

With the increasing cone angle, the gaps were enlarged by the conical plate so that the drag torque decreased. The smallest drag torque was 19.85 Nm, 37.73% smaller than the 0° group, present in the 1° case. The reason for this is that the thick oil film between the friction pairs was rendered ineffective, so the viscous nature of the oil was partly reduced. When the angle increased to 1.5°, since the reserved clearance was occupied mostly by the conical plates, the gaps at the outermost edge and innermost edge between the friction plates and separate plates decreased again. Accordingly, the oil film became thick and strong, and the drag torque increased again, i.e., higher than the 1° and 0.5° cases. In addition, though the 1° case displayed the smallest drag torque, the drag torque of 0.5° and 1.5° was obviously smaller than the original design. On occasions where the reserved clearance is too small to contain the 1° plates, the 0.5° plates may help. In applications where the lubricating oil has a higher viscosity, 1.5° plates are an alternative option, as they can provide a larger elastic restoring force.

Clutches are under separation status most of the time and thus, reducing the drag torque can conserve a large proportion of power and energy. Though the cost of the conical plate is higher and processing is more complicated, the cost and time needed for manufacturing are less than 10% more than those required for the flat plate. In mass production, the time and cost will be further decreased. In contrast, the effect of reducing the drag torque is clear. Therefore, by applying the conical plate, the drag torque is reduced significantly and much of the power can be saved.

5. Conclusions

To solve the incomplete separation problem and reduce the drag loss in wet multi-disc clutches, conical separate plates were designed. A numerical model, which evaluated the dynamic characteristics during the engagement and separation process, was verified via bench tests. No extra components are required, maintaining the compactness of the clutch. This investigation provides a theoretical basis for further studies on conical plates, clutch dynamics, and failure prevention. The results show that the advantages outweigh the disadvantages, especially in high-piston-pressure conditions. The main conclusions are as follows:

(1) The conical plate can provide elastic restoring force during the separation process to assist separation, reducing the separation time by 18.78%. The uniformity of the gaps is increased by 25.31%, resulting in a change in the lubrication status from the mixture lubrication to the hydrodynamic lubrication.
(2) The established numerical model can effectively describe the clutch engagement and separation characteristics, and its feasibility is verified by experiments. As evidenced by the simulations and experiments, the best layout is the 1° conical plates.
(3) By avoiding the sticking of the friction components, 37.73% of the drag torque is reduced and the transmitted torque is reduced by 10.31%. Since the separation status occupies most of the operating time, the trade-off is acceptable. In addition, under higher-piston-pressure conditions, the influence on the transmitted torque, but not the drag torque, can be reduced.

Author Contributions: Conceptualization, B.M. and C.X.; methodology, S.Y.; software, B.F.; validation, L.Y. and S.Y.; formal analysis, Q.Z.; investigation, Q.Z.; resources, B.M.; data curation, B.F.; writing—original draft preparation, Q.Z.; writing—review and editing, L.Y.; visualization, Q.Z.; supervision, C.X.; project administration, B.M.; funding acquisition, B.M., C.X., L.Y. and S.Y. All authors have read and agreed to the published version of the manuscript.

Funding: This research was funded by the National Natural Science Foundation of China, grant numbers 52205047, 52175037, and 51805289; the Frontier Cross Project of Beijing Institute of Technology, grant number 2024CX11006; and Shandong Provincial Natural Science Foundation, grant number ZR2024QE531. The APC was funded by the National Natural Science Foundation of China, grant number 51805289.

Data Availability Statement: The original contributions presented in this study are included in the article. Further inquiries can be directed to the corresponding author.

Conflicts of Interest: Author Shufa Yan was employed by the company Automotive Research Institute, China National Heavy Duty Truck Group. The remaining authors declare that the research was conducted in the absence of any commercial or financial relationships that could be construed as a potential conflict of interest.

References

1. Šabík, V.; Futáš, P.; Pribulová, A. Failure analysis of a clutch wheel for wind turbines with the use of casting process simulation. *Eng. Fail. Anal.* **2022**, *135*, 106159. [CrossRef]
2. Zhang, Z.; Mu, Z.; Yu, X. Mechanistic Study of Groove Parameters on the Thermoelastic Instability of Wet Clutch. *Lubricants* **2025**, *13*, 150. [CrossRef]
3. Wu, J.; Wang, L.; Li, L.; Shu, Y.; Yang, L.; Lei, T.J.M. Sliding State Analysis of Fractal Rough Interface Based on the Finite Element Method. *Materials* **2021**, *14*, 2121. [CrossRef] [PubMed]
4. Wu, Y.; Liu, Y.; Chen, H.; Chen, Y.; Xie, D. An investigation into the failure mechanism of severe abrasion of high-speed train brake discs on snowy days. *Eng. Fail. Anal.* **2019**, *101*, 121–134. [CrossRef]
5. Zhang, Z.; Zou, L.; Liu, H.; Feng, J.; Chen, Z. Response Characteristics of Dynamic Torque for Wet Clutch Engagement: A Numerical and Experimental Study. *Shock Vib.* **2021**, *2021*, 5522998. [CrossRef]
6. Pointner-Gabriel, L.; Schermer, E.; Schneider, T.; Stahl, K. Experimental analysis of oil flow and drag torque generation in disengaged wet clutches. *Sci. Rep.* **2023**, *13*, 17193. [CrossRef]
7. Falade, A.; Brenner, H. First-order wall curvature effects upon the Stokes resistance of a spherical particle moving in close proximity to a solid wall. *J. Fluid Mech.* **1988**, *193*, 533–568. [CrossRef]
8. Yuan, Y.; Liu, E.A.; Hill, J.; Zou, Q. An Improved Hydrodynamic Model for Open Wet Transmission Clutches. *J. Fluids Eng.* **2006**, *129*, 333–337. [CrossRef]
9. Jakubov, T.S.; Mainwaring, D.E. The surface tension of a solid at the solid–vacuum interface, an evaluation from adsorption and wall potential calculations. *J. Colloid Interface Sci.* **2007**, *307*, 477–480. [CrossRef]

10. Pointner-Gabriel, L.; Flamm, S.; Menzel, M.; Voelkel, K.; Stahl, K. Experimental investigation of drag loss and plate separation behavior of wet clutches under external forces. *Results Eng.* **2024**, *24*, 102918. [CrossRef]
11. Mahmud, S.F.; Pahlovy, S.A.; Kubota, M.; Ogawa, M.; Takakura, N. *Multi-Phase Simulation for Studying the Effect of Different Groove Profiles on the Drag Torque Characteristics of Transmission Wet Clutch*; 2016-01-1144; SAE International: Warrendale, PA, USA, 2016.
12. Fish, R.L. Using the SAE #2 Machine to Evaluate Wet Clutch Drag Losses. *SAE Trans.* **1991**, *100*, 1041–1054.
13. Pointner-Gabriel, L.; Menzel, M.; Voelkel, K.; Schneider, T.; Stahl, K. Experimental investigation of drag loss behavior of dip-lubricated wet clutches for building a data-driven prediction model. *Sci. Rep.* **2024**, *14*, 9241. [CrossRef] [PubMed]
14. Leister, R.; Najafi, A.F.; Gatti, D.; Kriegseis, J.; Frohnapfel, B. Non-dimensional characteristics of open wet clutches for advanced drag torque and aeration predictions. *Tribol. Int.* **2020**, *152*, 106442. [CrossRef]
15. Neupert, T.; Bartel, D. Evaluation of Various Shear-Thinning Models for Squalane Using Traction Measurements, TEHD and NEMD Simulations. *Lubricants* **2023**, *11*, 178. [CrossRef]
16. Hilpert, C.R. Gyroscopically Induced Failure in Multiple Disc Clutches, Its Causes, Its Characteristics and Its Cures. *SAE Trans.* **1969**, *78*, 354–371.
17. Hou, S.; Hu, J.; Peng, Z. Experimental Investigation on Unstable Vibration Characteristics of Plates and Drag Torque in Open Multiplate Wet Clutch at High Circumferential Speed. *J. Fluids Eng.* **2017**, *139*, 111103. [CrossRef]
18. Yu, Y.; Peng, Y.; Lan, S.; Zhou, P. Finite Element Analysis for Wave Spring of Multi-Disc Wet Clutch Based on ANSYS. *Appl. Mech. Mater.* **2013**, *419*, 203–208. [CrossRef]
19. Erfanian-Naziftoosi, H.R.; Shams, S.S.; Elhajjar, R. Composite wave springs: Theory and design. *Mater. Des.* **2016**, *95*, 48–53. [CrossRef]
20. Li, S.-X.; Liao, H.-R.; Li, S.-C.; Chen, L.; Li, Q.-Z.; Cheng, T.-F.; Chen, K.-Y. Effect of structural parameters on the bearing characteristics of wave spring. *J. Mach. Des.* **2020**, *37*, 22–27. [CrossRef]
21. Yan, Z.; Wu, B.; Gao, X.; Bai, X. Engagement/Disengagement Characteristics of Pull-Type Diaphragm Spring Clutch for Heavy-Duty Commercial Vehicles. *Chin. J. Mech. Eng.* **2024**, *37*, 124. [CrossRef]
22. Wang, H.; Zhu, C.; Yang, X.; Niu, Y.; Bao, H.; Zhu, R. Study on the engagement characteristics of wet multi-plate friction clutch based on separating spring. *Forsch. Im Ingenieurwesen* **2024**, *88*, 3. [CrossRef]
23. Wu, J.; Cui, J.; Shu, W.; Wang, L.; Li, H. Coupling mechanism and data-driven approaches for high power wet clutch torque modeling and analysis. *Tribol. Int.* **2024**, *191*, 109166. [CrossRef]
24. Zhao, Q.; Ma, B.; Yu, L.; Zhang, Y.; Dong, Y.; Zhou, R. Numerical and experimental studies on the thermodynamic characteristics of post-buckling separate plate in the clutch. *Tribol. Int.* **2024**, *195*, 109607. [CrossRef]
25. Matsuoka, T. Study on Effect of Oiliness Agents in the Presence of Detergents on Friction Characteristics of Paper-Based Friction Materials Under Low Sliding Velocity. *Jpn. J. Tribol.* **2002**, *47*, 488–497.
26. Greenwood, J.; Williamson, J. Contact of Nominally Flat Surfaces. *Proc. R. Soc. Lond.* **1966**, *295*, 300–319. [CrossRef]
27. Jung, W.; Hyeon, J.; Doh, N. Robust Cuboid Modeling from Noisy and Incomplete 3D Point Clouds Using Gaussian Mixture Model. *Remote Sens.* **2022**, *14*, 5035. [CrossRef]
28. Bao, H.; Xu, T.; Jin, G.; Huang, W. Analysis of Dynamic Engaged Characteristics of Wet Clutch in Variable Speed Transmission of a Helicopter. *Processes* **2020**, *8*, 1474. [CrossRef]
29. Zheng, L.; Ma, B.; Chen, M.; Yu, L.; Xue, J. Numerical studies on the dynamic characteristics of a wet multi-disc clutch during the disengaging process. *J. Mech. Sci. Technol.* **2022**, *36*, 3277–3289. [CrossRef]
30. Shutin, D.; Kazakov, Y. Theoretical and Numerical Investigation of Reduction of Viscous Friction in Circular and Non-Circular Journal Bearings Using Active Lubrication. *Lubricants* **2023**, *11*, 218. [CrossRef]
31. Yu, L.; Ma, B.; Chen, M.; Li, H.; Liu, J.; Li, M. Investigation on the failure mechanism and safety mechanical-thermal boundary of a multi-disc clutch. *Eng. Fail. Anal.* **2019**, *103*, 319–334. [CrossRef]

lubricants

MDPI

Article

Tribo-Electric Performance of Nano-Enhanced Palm Oil-Based Glycerol Grease for Electric Vehicle Bearings

Amany A. Abozeid [1], May M. Youssef [1], Tamer F. Megahed [2,3], Mostafa El-Helaly [1], Florian Pape [4,*] and Mohamed G. A. Nassef [1,5,*]

1 Production Engineering Department, Faculty of Engineering, Alexandria University, Alexandria 21544, Egypt; eng.amany.abozeid@alexu.edu.eg (A.A.A.); may_youssif@alexu.edu.eg (M.M.Y.); mahelaly@alexu.edu.eg (M.E.-H.)
2 Electrical Power Engineering Department, Egypt-Japan University of Science and Technology (E-JUST), Alexandria 21934, Egypt; tamer.megahed@ejust.edu.eg
3 Electrical Engineering Department, Mansoura University, El-Mansoura 35516, Egypt
4 Institute of Machine Design and Tribology, Leibniz University, 30823 Hanover, Germany
5 Industrial and Manufacturing Engineering Department, Egypt-Japan University of Science and Technology (E-JUST), Alexandria 21934, Egypt
* Correspondence: pape@imkt.uni-hannover.de (F.P.); mohamed.nassef@ejust.edu.eg (M.G.A.N.)

Abstract: Rolling Bearings are crucial components for induction motors and generators in electric vehicles (EVs), as their performance considerably influences the system's operational reliability and safety. However, the commercial greases used for bearing lubrication in EV motors pose a detrimental impact on the environment. In addition, they are ineffective in mitigating the effect of electric discharges on rolling surfaces leading to premature bearing failures. This study investigates the viability of a developed eco-friendly grease from palm olein as the base oil and glycerol monostearate as the thickener, enhanced with conductive multi-walled carbon nanotubes (MWCNTs) for EV motor bearings prone to electrical currents. Chemical–physical, tribological, and electrical tests were conducted on the developed grease samples without and with MWCNTs at 1 wt.%, 2 wt.%. and 3 wt.% concentrations and results were compared to lithium and sodium greases. Palm grease samples demonstrated a lower EDM voltage range reaching 1.0–2.2 V in case of 3 wt.% MWCNTs blends, indicating better electrical conductivity and protecting the bearing surfaces from electric-related faults. These findings were further confirmed using vibrations measurement and SEM-EDX analysis of the electrically worn bearings. Bearings lubricated with palm grease blends exhibited lower vibration levels. Palm grease with 2 wt.% MWCNTs reduced vibration amplitudes by 28.4% (vertical) and 32.3% (horizontal). Analysis of bearing damaged surfaces revealed enhanced damaged surface morphology for MWCNT-enhanced palm grease as compared to surface lubricated by commercial greases. The results of this work indicate that the proposed bio-grease is a promising candidate for future application in the field of next-generation electric mobility systems.

Keywords: rolling bearings; induction motors; breakdown voltage; capacitance EDM currents; multi-walled carbon nanotubes; green lubricants

1. Introduction

Electric vehicles (EVs) are recently introduced to the international market as a sustainable and eco-friendly mobility system in comparison with traditional internal combustion engine (ICE) vehicles. Due to technological advancements, increased environmental awareness, and supportive governmental policies [1–4], EVs are gaining broad acceptance and are increasingly viewed as a viable and desirable alternative to reduce greenhouse gas emissions and the environmental effect of transportation [5]. They possess excellent benefits, such as high reliability, high power density, high efficiency, and the capability of instant start [6]. The market share of EVs has been increasing, with countries such as China, the EU, and the United States leading the change from IC engine vehicles to electric alternatives in 2023 alone, with almost 14 million [7]. In 2023, electric vehicles were registered worldwide, recording an increase of 35% from 2022 and reaching around 40 million operational units [6]. The European Environment Agency reports that the harmful gas emissions from EVs during their lifecycle are up to 30% lower than those of ICE vehicles. It is expected that the lifecycle emissions of a typical electric vehicle could be reduced by at least 73% by 2050 [8].

EVs are equipped with almost the same motor drive system consisting of a DC/DC converter for voltage regulation, an inverter for motor control, and an electric motor. Much of the stability in these motor systems contributes to the reliability of the entire EV. Electric vehicles use various types of motors: DC motors, induction motors, permanent magnet motors, PM brushless DC motors, and switched reluctance motors [9,10]. Among these, the three-phase induction motor remains the most-used prime mover [11,12]. Yet, parasitic currents problem arises in EVs from the inherent resistances, inductances, and capacitances in components like DC-DC converters and Pulse Width Modulation (PWM) inverters [13,14]. PWM technique in variable frequency drives (VFDs), commonly used in electric vehicle (EV) inverters to control motor speed and torque, can generate high-frequency switching events that induce common-mode voltages on the motor stator windings [15,16]. These rapid voltage changes, combined with the inherent resistances, inductances, and especially the parasitic capacitances between motor windings, cables, and the vehicle chassis, result in the buildup of shaft voltages. The parasitic capacitances form an electrical pathway that allows capacitive currents to flow from the stator through the rotor and along the shaft. In this case, the lubricant used for the bearings acts as an insulator, allowing electrical charge to accumulate. When the shaft voltage across the bearing exceeds the dielectric strength of the lubricant film, the oil film can no longer insulate the bearing elements, leading to transient breakdown events. This causes a sudden discharge of current through the bearing—known as electric discharge machining (EDM) current—which passes through the thin lubricant film in the form of miniature electric arcs.

Many researchers have taken up extensive research work to explain the relationships between different influential parameters, mechanisms, and corresponding modes of failures that affect the motor bearings [17–19]. EDM-bearing currents occur through two main mechanisms. Under a fully lubricated condition, the insulating properties of the bearing lubricant allow it to act as a capacitor. EDM currents are generated when the common-mode voltage surpasses the dielectric strength of the lubricant, resulting in electrical discharges within the bearing [20]. Under such conditions, the lubricant shifts instantaneously from behaving as a capacitor to functioning as a resistor. Additionally, EDM currents can occur when the lubricant film fails to adequately separate the bearing surfaces, leading to contact between surface asperities [21]. This breakdown of the lubricant film creates short circuits, triggering electrostatic discharges. These currents can range between 0.5 and 3 A. Since their magnitude is generally unaffected by motor size, EDM currents pose a greater risk to smaller motors (less than 110 kW) [22,23]. The rapid energy dissipation from these discharges can cause localized temperature spikes, leading to surface melting in the bearings. Such melting often results in microscopic craters, as well as larger-scale frosting, fluting damage, white etching cracks, and lubricant degradation [24].

The common methods for mitigating this problem have been insufficient to eliminate bearing currents, and the mechanism of how these currents damage the bearings is still not well understood [25]. Several mitigation techniques have been devised to solve this problem. For instance, shaft grounding brushes are meant to create a low-impedance path for stray currents, rerouting them away from bearings. These brushes, however, need constant maintenance and replacement because of wear and contamination. Neglecting routine maintenance can lead to high impedance, thereby diminishing their performance over time [26]. Insulated bearings are made in such a way that current passage is avoided through the use of non-conductive materials [27] and coatings [28]. Despite this design, insulation breakdown can occur due to inherent material defects or presence of high-frequency voltages that allow current passage and lead to damage.

While many attempts are made to eliminate them, shaft voltages cannot be completely avoided [29]; thus, studies on their effects become more important in electric and hybrid electric vehicles. When electric discharges flow through the bearing elements, it creates microscopic arcing damage on the raceway surface. When rolling elements pass over these damaged areas, this generates vibrations that result in a thinner lubricant film, which in turn enables easier passages of subsequent current. As time passes, this process creates grooves or holes on the raceway, which gradually raise vibration levels, initiate lubrication breakdowns, leading to instability, and noise, which eventually leads to bearing mechanical failure [30,31].

The lubrication failure in electrically charged scenarios has been a frequent concern due to the altering of the chemical composition and heating of the lubricant on a localized scale by electric sparks [32,33]. The energy dissipation forms conductive particles in the lubricant, depletes critical additives, and promotes oxidation and sludge formation [34]. The presence of these particles adds more electrical pathways, increasing the chance of discharges. In addition, wear of the lubricant film reduces its capacity to keep contact surfaces apart, resulting in more friction and wear, and an increased risk of bearing mechanical failure [35–37]. Hence, parasitic currents have been reported to shorten the useful life of lubricants drastically, requiring more frequent maintenance and creating a reliability challenge for electric vehicles and industrial motors [38,39].

Significant efforts have been devoted to improving bearing lubricants to reduce the impact of parasitic currents. Incorporating superior additives—antioxidants, anti-wear agents, and extreme-pressure enhancers—has raised the thermal and chemical stability level in greases, even when they are subjected to rigorous working environments [34]. Nanomaterials such as activated carbon nanoparticles [38,39], carbon nanotubes, boron nitride, graphene [40], and graphene oxide have been recently found to be environmentally friendly anti-wear (AW) and extreme-pressure (EP) additives for lubricants [41,42]. The addition of nano-additives has shown promise in enhancing tribological and physical performance in lubricants [43]. Furthermore, current research focuses on the development and investigation of conductive greases formulated with ionic liquids [44]. Despite such developments, there are still insufficient studies that explored the application of advanced metallic and carbon-based nano-additives to enhance the tribo-electric behavior of bearing greases such as silver nanoparticles [45], alumina (Al2O3), and multi-walled carbon nanotubes (MWCNTs) [46].

Traditional lubricants are usually fossil fuel-based, thus highly contributing to high carbon emissions, harm to the environment due to poor disposal methods, and biodegradability problems, all of which pose significant challenges to environmental sustainability [47]. Bio-based lubricants derived from renewable sources present themselves as an alternative solution that, in addition to decreasing dependence on limited fossil fuel resources, also demonstrate improved viscosity, lower toxicity, and reduced greenhouse gas emissions during their lifecycle [48]. Vegetable oils, such as rapeseed oil [49], Soybean oil [50], Jatropha oil [51], and Cottonseed oil [52] have been receiving significant interest because of the better biodegradability compared to animal fats and esters [53]. It has also been recently demonstrated that lubricating greases based on vegetable oils provide an adequate and homogeneous full lubricant film with comparable load-carrying capacity to conventional lithium greases [54,55]. Palm oil has emerged as a potential alternative to traditional mineral oils as a lubricant due to its distinguished physical properties and high production efficiency [56]. It is composed of free fatty acids like linear carboxylic acids with carbon chains between 12% and 24%. Its main components like oleic (monounsaturated), linoleic (polyunsaturated), palmitic, and stearic acids are responsible for the efficiency of palm oil as a lubricant to minimize wear and friction [57,58].

Few tribological studies [59–61] have evaluated palm olein blended with mineral oils for reciprocating equipment. Nevertheless, palm oil has drawbacks concerning low oxidation stability and marginal kinematic viscosity values, further limiting its applicability in mechanical systems. Hence, researchers have tested various additives to overcome the drawbacks [62].

Based on the previous review of the literature, most of the papers focused on the impact of electrical discharge machining (EDM) currents on bearing degradation. They explored different mitigation strategies such as shaft grounding brushes, insulated bearings, and improvements in lubricant properties. These studies have highlighted the importance of addressing parasitic currents and their detrimental effects on bearing life, lubricant degradation, and motor performance in electric vehicles (EVs). However, only a few research works have addressed the impact of applying nano-additives to bearing grease on the severity and mode of damage in bearing elements due to electric faults. Furthermore, the literature lacks any effort to consider replacing conventional lithium grease with more environmentally friendly and sustainable lubricants in EV bearings as a proposed solution to mitigate or control the parasitic current damage effect.

This work investigates the effect of a palm oil-derived glycerol bio-grease on bearing tribo-electric performance under parasitic electrical currents. The emphasis is on assessing bearing electrical degradation under stepwise DC voltages (0–10 V) with palm grease in plain form and with varying concentrations of MWCNT additives. A custom test setup with an electrical circuit across the bearing rings is employed. Voltages as low as 5 V can lead to electrical bearing damage, especially in systems equipped with variable frequency drives (VFDs), due to parasitic currents and electrical discharge machining (EDM) effects [63,64]. Although AC and DC discharge characteristics are different, dielectric breakdown and EDM-induced damage mechanisms are similar. DC voltage testing here is employed to determine the threshold for EDM events and evaluate each lubricant's ability to suppress electrical damage. Voltage and current signals are measured using sensors and an oscilloscope to characterize capacitive, EDM, and resistive responses. Bearing damage severity is assessed by vibration analysis under 100 N radial load and 1400 rpm, and surface analysis is conducted through SEM/EDX to confirm the protective role of palm grease and MWCNTs. The findings are aimed at facilitating the development of sustainable, reliable lubricants for EV applications. The manuscript is organized as follows:

Section 2—details of the test procedures, conditions, and six grease samples: (A) lithium, (B) sodium, (C) palm grease without additives, (D) palm grease with 1 wt.% MWCNTs, (E) palm grease with 2 wt.% MWCNTs, and (F) palm grease with 3 wt.% MWCNTs.

Section 3—comparative analysis of EDM damage mitigation.

Section 4—summary of findings and implications for EVs sustainable lubrication.

2. Materials and Methods

2.1. Test Rig

A customized test rig was designed and fabricated to conduct performance evaluations of ball bearings lubricated with various palm grease blends, as shown in Figure 1. The rig consists of an induction motor, flexible coupling, shaft, two supporting deep-groove ball bearings, and the test bearing mounted on a specially machined hub. A vertical radial load of 100 N is applied to the test bearing through a hinged loading mechanism that ensures constant contact. The radial load of 100 N used in this study was chosen by carefully balancing mechanical test rig constraints and relevance to actual electric vehicle (EV) motor bearing operating conditions.

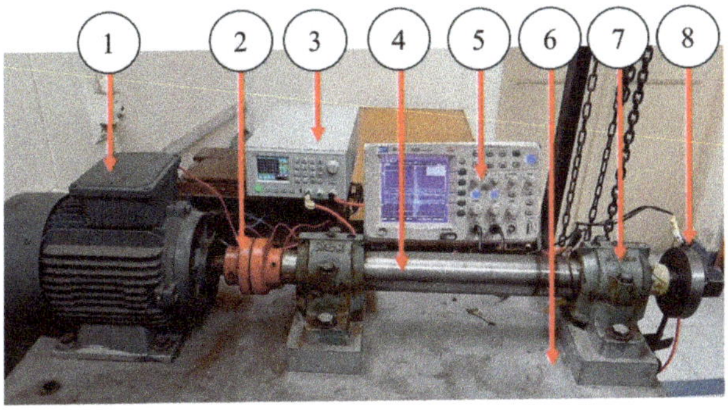

Figure 1. The assembly model of the test rig setup consists of (1) an induction motor, (2) coupling, (3) a DC power supply, (4) a shaft, (5) a digital oscilloscope, (6) a base, (7) a supporting bearing, and (8) a special hub of the test bearing with an applied radial load vertically hinged to it.

Firstly, the test setup features an overhung spindle with the 6006ZZ deep groove ball bearing positioned at the shaft end and loaded via a simple hanger mechanism attached to the bearing housing. Initial trials indicated that radial loads ranging from 50 N to 300 N enable a stable bearing operation without excessive spindle deflection or misalignment, thus avoiding undesirable vibration and preserving experimental repeatability. A value of 100 N was selected within this stable range. Secondly, the 100 N load is well below the dynamic load rating (13,200 N) of the 6006ZZ bearing tested (as listed in Table 1), consistent with SKF catalog recommendations for moderate-duty applications. This load represents a realistic operating condition for medium-sized electric motors used in passenger EVs, where typical radial bearing loads vary widely depending on design and service conditions. According to reference [65], typical radial loads in electric motors commonly range from 100 N to 1000 N. The selected 100 N load thus reflects the lower bound of realistic radial loading encountered in motor bearings, providing a relevant and conservative test condition for tribological and electrical discharge evaluation. This configuration enables stable operation at a fixed rotational speed of 1400 rpm and operating temperatures maintained below 60 °C.

Table 1. Specifications of the test bearing (6006zz) provided by the NSK manufacturer.

Dimensions (mm)	d	D	B	r
	30	55	13	1
Mass (kg)	0.116			
Dynamic load rating, C (N)	13,200			
Static load rating, Co (N)	8300			
Clearances (μm)	C3 (13–28)			

Additionally, an electrical circuit was developed across the test bearing hub using a programmable DC power supply, a digital storage oscilloscope, and precision electric terminals to monitor voltage and current across the bearing. This setup enables controlled application of DC voltages (0–10 V), with a current limit of 3 A, mimicking shaft voltage buildup and bearing current flow scenarios typically encountered in inverter-fed induction motors for EVs. Voltage increments of 0.5 V were applied every 5 min, and the onset of discharge events, indicating electrical discharge machining (EDM), was recorded to identify the transition from capacitive to resistive regimes.

While this laboratory setup does not fully reproduce the dynamic and thermally fluctuating conditions of real-world EV systems, the selected operating parameters were carefully aligned with reported ranges for electric motor bearing environments in EV applications, particularly with regard to electrical stress and moderate mechanical load [27,66,67].

The power supply is used to generate the desired DC voltage across the test bearing hub using electric connectors and wiring installed into the hub, where one terminal is to be connected to the bearing outer ring while the other terminal is attached to the shaft and inner ring. The inner ring of the test bearing is electrically insulated from the bearing hub using a polyethylene film. A two-channel digital oscilloscope is exploited to monitor and record both the applied voltage and current during the electric tests, where the first channel is used to record the voltage signal while the second channel is connected to a current sensor that monitors the current signal across the bearing raceways.

It is acknowledged that the electrical environment in electric vehicle bearings is dominated by AC and high-frequency PWM voltages, which can result in more sever discharge behaviors and surface damage mechanisms than those produced by DC excitation. In this study, DC voltage was employed to enable controlled, stepwise assessment of the dielectric breakdown and electrical discharge mitigation performance of the tested lubricants. This approach allows for direct comparison of the intrinsic electrical insulating properties of each grease formulation. DC testing simplifies the electrical environment, allowing for accurate detection of leakage current, electrical discharge machining (EDM) thresholds, and tribofilm formation without the complexities of alternating polarity and high frequency switching transients. This approach is consistent with methodologies reported in prior research aimed at isolating and benchmarking the electrical behavior of bearing lubrication systems [68–70].

The characteristics of the setup components are listed in Table 2. The selected bearing for this investigation is a deep-groove ball bearing of type NSK 6006ZZ (NSK distributer—Arabian Co for Electrical & Mechanical Supplies, Alexandria, Egypt) to imitate the used bearing type in the EV induction motors [71]; their main information are listed in Table 2 according to NSK bearing manufacturer (NSK Ltd., Tokyo, Japan) [72]. In this work, the motor runs at 1400 RPM while a dead weight of 100 N is hinged to the bearing hub through a suitable hook.

Table 2. Specifications of test rig components.

Component	Specifications
Electric Motor	GAMAK (3 hp, and 1400 rpm)
Base	C45 Carbon Steel
Shaft	SUS 420 Stainless Steel
Two Support Bearings	NU1011M Roller Bearing
Dc power supply	RD6024/RD6024-W
Digital oscilloscope	DQ7022S, two-channel (25 MHz)

2.2. Characterization of Nanomaterials

Multi-walled carbon nanotubes (MWCNTs) with a diameter of 10–40 nm and a length of up to 5 μm are selected in this work as a carbonaceous nano-additive material with anisotropic characteristics. Based on the local manufacturer product specifications, they possess a purity level of 92 ± 2%.

To analyze the structural characteristics of MWCNTs, including particle dimension, particle morphology, and the amount of particles clustering, a scanning electron microscope (SEM) (JEOL JSM-IT200, JEOL Ltd., Tokyo, Japan) and a transmission electron microscope (TEM) (JEOL JEM-2100, JEOL Ltd., Tokyo, Japan) were employed. The sample preparation involved adhering the powder sample to double-sided carbon tape, then coating the sample with a platinum–palladium layer at a current of 40 mA using the JEOL JEC-3000 FC Auto Fine Coater. For TEM analysis, the preparation process included dispersing the MWCNT powder in ethanol and subjecting the mixture to ultrasonic agitation for 15 min.

2.3. Grease Synthesis

The proposed lubricant in this work is manufactured using 70% palm olein (base oil) and 30% glycerol monostearate (thickener). The preparation process is planned to ensure proper mixing, temperature control, and homogenization to achieve the desired grease properties [73,74]. Initially, a 2 L beaker filled with palm olein is placed on a magnetic stirrer with a heating function and heated to a temperature of 120 °C. The heat treatment in this step reduces palm olein's viscosity, hence efficiently allowing the incorporation of the thickener.

As the palm olein reaches the target temperature, 20% glycerol is gradually added to the beaker while maintaining continuous stirring to ensure uniform distribution within the palm olein matrix. The temperature is sustained within 120 °C for 30 min, allowing sufficient interaction and partial integration of the components. An additional stirring period of 15 min ensures homogeneity [75]. After the initial heating and mixing, the flask is kept at 120 °C, and the remaining 10% of the glycerol is added to the mixture. This addition is critical for achieving the desired consistency and further blending the components. The mixture is stirred at 120 °C, so the new olein spreads uniformly [76]. Finally, it is left to cool to room temperature before mixing it with a commercial-grade mixer for 30 min. It will ensure the glycerol is evenly dispersed in the palm olein, creating stable and well-structured grease.

The mixing step is conducted utilizing a mechanical mixer until the output grease appearance is uniform and its color is milky yellow in the case of palm grease in its plain form and black in the cases of samples of palm grease with three different blends of 1 wt.%, 2 wt.%, and 3 wt.% MWCNTs, respectively. Samples of each blend are weighed and applied to the identical test ball bearings guided by SKF formula in Equation (1) [42].

$$G_q = 0.005(D)(B) \tag{1}$$

where Gq is the grease quantity to lubricate the bearing (g), D is the outside diameter of the ball bearing (mm), and B indicates the bearing width (mm). Palm grease samples with the selected concentrations of multi-walled carbon nanotube (MWCNT) powder are prepared by incorporating the MWCNTs into the base grease matrix through mechanical stirring. The mixing process is carried out using a laboratory-grade mechanical stirrer to ensure homogeneous dispersion of the nanoparticles throughout the grease medium. Following preparation, the samples are labeled according to their respective MWCNT concentrations and subsequently subjected to a comprehensive series of evaluations, including chemical, physical, tribological, and electrical characterization, as summarized in Table 3. For benchmarking purposes, conventional lithium grease and sodium grease—each composed of

80% mineral oil and 20% thickener—are tested under identical experimental conditions. The subsequent sections detail the methodologies and procedures employed to assess the performance characteristics of the formulated palm-based greases.

Table 3. The developed test grease blends.

Grease Sample Label	Grease Blends
Palm Grease A	Palm-oil-based glycerol grease (in its plain form)
Palm Grease B	Palm grease with 1 wt.% MWCNTs
Palm Grease C	Palm grease with 2 wt.% MWCNTs
Palm Grease D	Palm grease with 3 wt.% MWCNTs

2.4. Kinematic Viscosity Test

The palm oil is tested for its kinematic viscosity at 40 °C and 100 °C, according to ASTM-D445 [77]. For this purpose, a sample volume of 25–30 mL is forced to flow in a capillary tube of a viscosity meter surrounded by a temperature-controlled liquid bath. The test is repeated three times, and the average value of kinematic viscosity is recorded. Based on the literature review, the recommended ISO VG (International Organization for Standardization Viscosity Grade) for ball bearings falls between VG 68 and VG100 [78].

2.5. Dropping Point Test

To evaluate the palm grease consistency at elevated temperature in rolling bearings during operation, the dropping point is determined according to ASTM D2265 [79,80]. A 10 g sample of the prepared grease is placed in a brass cup, supported by a glass test tube, and placed in an aluminum block oven. The temperature gradually increased to 232 °C in the oven, with the temperature observed using an installed thermometer to the oven. Once an oil droplet falls from the cup into the test tube, the temperature is recorded by the thermometer and identified as the grease's dropping point, indicating its heat resilience.

2.6. Thickener Consistency and Oil Bleeding Test

The consistency of the developed grease is evaluated to determine the resistance of the fibrous structure of the grease to a specified radial load applied in a standardized manner. A grease with adequate consistency ensures that the grease thickener possesses the required stiffness during service, while allowing base oil to be released through structure channels for lubrication in machinery components. A cone penetration test is used for this purpose according to ASTM D217 [62,78]. A metallic cup is completely filled with a grease sample size of 500 g. Then, a penetrometer of a cone shaped load is set free to drop vertically into the cup with the sample and is left for 5 s at a temperature of 25 ± 5 °C before recording the depth reading on a dial gauge in tenths of a millimeter. The output reading is compared to the National Lubricating Grease Institute (NLGI) scale to identify the consistency of the grease.

The static bleeding test is performed according to ASTM D1742 [81]. Each grease sample is placed in a sieve positioned above an aluminum crucible, with a static load of 100 g applied directly to the grease. The sample is then placed in an oven at 40 °C for 18 h. After the test period, the released oil is collected, weighed, and the amount of oil separation is calculated by dividing the weight of the released oil by the total amount of grease.

2.7. Tribological Test

The friction coefficient of palm grease samples was measured using the ball-on-disk method according to ASTM G99 [82]. The test method is designed to simulate sliding/rolling contact between lubricated surfaces with the aim of studying the frictional behavior of the grease under controlled conditions. A steel ball 10 mm in diameter rests on the surface of a rotating disk with a diameter of 200 mm as described in Figure 2. A thin, uniform layer of grease (30–40 mg) is evenly distributed over the whole disk surface prior to each test using lint-free cloth. Though ASTM G99 [82] does not mandate precise mass, we ensured consistent coverage by measuring and distributing the grease so that no dry spots formed. The disk rotates at a uniform speed, normally 600 rpm, while the rolling element (ball) is pressed against the disk under some controlled radial load of 10 N for a sliding distance of 1000 m of sliding distance. The high viscosity and consistency of the palm grease (NLGI 4) and its thickener structure (glycerol monostearate) ensured stable adhesion to the disk surface under these moderate speeds. This stability was further confirmed by the repeatability of the stable coefficient of friction (COF) measurements across three trials for each sample.

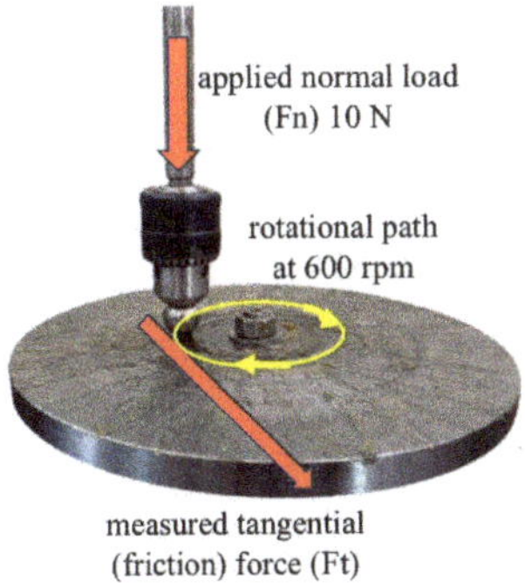

Figure 2. Pin-on-disk test setup.

This test is performed at 25 °C to maintain consistency in the result. During testing, the friction force is generated at the contact point between the stationary pin and the rotating disk due to the resistance to relative motion. This force acts tangentially to the circular path of the pin and in the direction opposite to the surface velocity of the rotating disk. Its magnitude is proportional to the applied normal load and the coefficient of friction between the two surfaces. The friction force is instantaneously measured using a load cell attached to the pin holder. The measured frictional force is divided by the applied normal load to obtain the COF. Following every test run, the average COF will be calculated over each condition.

2.8. Electric Tests

One of the main objectives of this study is to identify the voltage ranges across the lubricant grease-separating bearing surfaces in which capacitive, electric discharge machining (EDM), and resistive currents occur. This is done to examine how the conductivity of the lubricants influences the occurrence and extent of harmful EDM-related phenomena [83,84]. To achieve this, the electrical tests are carried out in three sequential phases, as explained in Figure 3. The first phase, referred to as the run-in stage, involves continuously operating each test bearing for 16 hrs. A moderate radial load of 100 N is applied to the bearing hub, and the induction motor is set to run at a constant speed of 1400 rpm, as demonstrated in Figure 4. This stage aims to smooth out surface asperities, smoothing out the bearing raceways and rolling elements. Additionally, this step helps in reaching a full film lubrication condition, preventing direct contact between asperities that could interfere with voltage buildup [85,86]. After this stage, the electrical test sequence begins with an incremental voltage increase from 0 to 10 V in 0.5 V steps every 5 min to determine the capacitive regime and the threshold of EDM currents. Once the EDM voltage is reached, the test continues for approximately 6 h to ensure sufficient time for bearing exposure to electric discharges initiating possible damage mechanisms. By the end of this phase, the voltage increments are increased by additional 0.5 V steps every 5 min until resistive current regime is reached and confirmed by the oscilloscope monitoring of voltage and current signals. The entire test was conducted at a controlled room temperature of 25 °C. The lubricant can be viewed as a parallel RC (resistor–capacitor) circuit—accurately represents the basic electrical behavior of a lubricated rolling bearing under applied voltage, according to the literature [63,64]. When the lubricant film is intact, it separates the bearing's rolling elements and races, acting as a dielectric between two conductors acting as a capacitor (C). When the voltage exceeds the dielectric strength of the lubricant film, the film breaks down and allows current to flow a resistive path form (R). Hence, the two paths (capacitive and resistive) exist simultaneously. Under normal (non-breakdown) conditions, only the capacitive path is active. During breakdown, the resistive path is activated, and both may be conducted (especially during transient events).

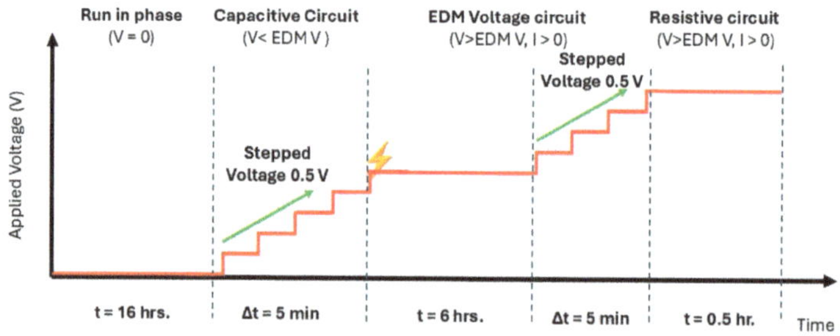

Figure 3. The planned procedure for conducting electric test runs on each test bearing.

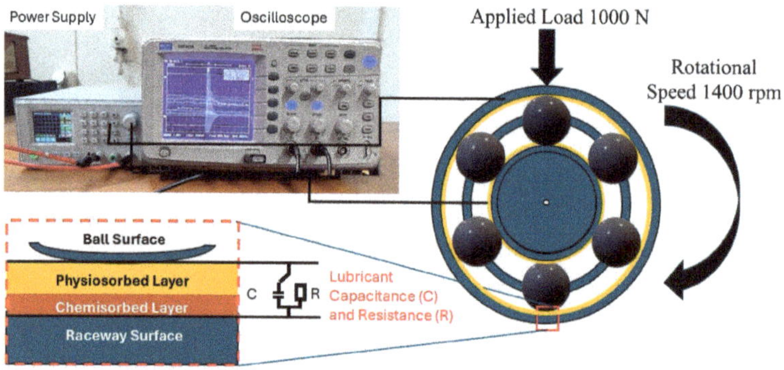

Figure 4. A schematic representation of the developed electric circuit including the ball bearing with wire terminals attached to the outer race and inner ring and connected to a power supply.

In the second phase, a DC voltage starting from 0 V and increased incrementally up to 10 V is applied to the bearing–lubricant system, with the current limit set to 3 A, values consistent with those reported in previous studies [87,88]. At the onset of this phase, the voltage is raised in 0.5 V increments every 5 min. This controlled increase enables careful monitoring of the electrical behavior across the bearing, particularly for identifying the characteristic domains of capacitive, electrical discharge machining (EDM), and resistive responses, as observed through a digital oscilloscope.

Initially, the electrical current remains negligible or near zero due to the capacitive nature of the system. However, once the applied voltage reaches a threshold sufficient to initiate EDM, sharp transient discharge events are detected. At this point, a sudden increase in current is typically observed, marking the onset of breakdown and conductive pathways through the lubricant film or surface asperities. The voltage at which these discharge events first occur is recorded as the EDM threshold voltage. Following this identification, the bearing is operated for 6 h under this constant potential difference to investigate discharge-induced degradation. Subsequently, the incremental voltage increase continues until the electrical behavior transitions into a stable resistive regime, where conduction becomes continuous rather than discharge-based, aligning with findings in prior literature [87].

2.9. Vibrations Analysis Test

The dynamic performance of each test bearing is evaluated after applying electric tests using vibration measurement and analysis as shown in Figure 5. This step is essential to stand upon the level and type of damage that took place in each electrocuted bearing, and how each grease blend lessened the harmful effect of parasitic currents on the surfaces of each bearing element. At this stage of work, the setup is set to run while the test bearing is still installed inside the hub after being disconnected from the electric circuit. Bearing vibration signals are measured using a uniaxial accelerometer mounted on the bearing hub by a magnet. The vibration signals are collected in both vertical and horizontal directions for analysis using a 2-channel data analyzer (Commtest VB5, Bentley Nevada, NV, USA). Vibration measurements are processed using special filters and analyzed in time domain and frequency domain. The amplitude of vibration levels is displayed as acceleration (mm/s^2), and the mean and standard deviation of three consecutive spectra results for the same test bearing are calculated.

Figure 5. Vibration analysis using commtest VB5 data collector and accelerometer mounted in radial direction to the test bearings.

3. Results and Discussion

3.1. SEM and TEM Results

Figure 6 demonstrates the SEM and TEM results of MWCNT powder. Based on the examination of Figure 6a, the MWCNTs exhibit an entangled and intertwined rope-shaped structure in bundles. The TEM image in Figure 6b further confirms their curved hollow tube structures, which are typical of MWCNTs. The nanotubes have average diameters ranging from 10 to 15 nm, with lengths extending up to 5 μm. According to the supplier, the purity percentage of the supplied MWCNTs is around 92 ± 2%, which is confirmed by the SEM and TEM observations.

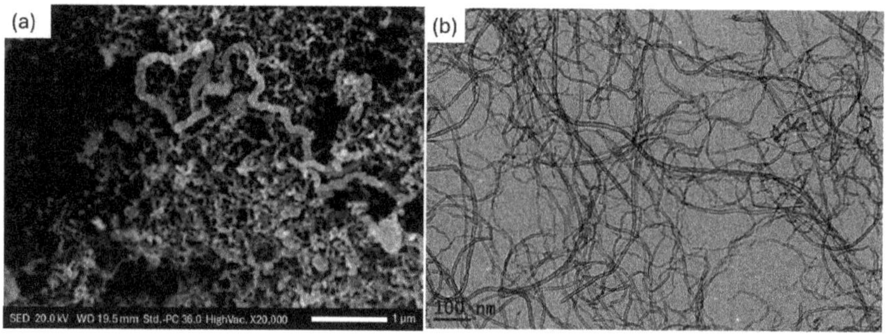

Figure 6. Microscopic examination results of the MWCNT powder using (**a**) SEM and (**b**) TEM.

3.2. *Physical–Chemical Results*

The developed bio-grease is tested for its physical–chemical characteristics and compared with commercial lithium and sodium greases, as presented in Table 4.

The grease sample of palm oil shows an NLGI 4 with a relevant unworked penetration value of 185 dmm. This penetration value is considerably lower than the two benchmark commercial greases, indicating the higher rigidity of the palm oil-based formulation. Regarding the bleeding properties, the bio-grease sample demonstrates a higher percentage of oil release, compared to commercial greases [89]. This is illustrated by the sensitivity of the glycerol thickener consistency in particular to the temperature, which leads to the breakage of structural bonding, yielding more oil separation [90]. While the static oil bleeding rate of >6% is higher than the normal level for conventional applications, this characteristic should be evaluated within the specific context of EV bearing requirements. The higher oil release rate can be beneficial in electrically stressed applications, as it ensures continuous replenishment of the lubricating film at the contact points where electrical discharges occur. Furthermore, modern EV designs often incorporate sealed bearing units with controlled environments, where oil bleeding is less problematic. The trade-off between some conventional performance metrics and the environmental and electrical benefits represents a deliberate design choice that aligns with the growing emphasis on sustainability in the automotive sector.

Palm grease was synthesized using 30% glycerol monostearate as a thickener, which forms a robust crystalline network that traps palm olein (base oil). This proportion of glycerol in the formulation targeted increasing the dropping point and mechanical stability of the grease, allowing it to maintain its consistency well beyond 40 °C. The dropping point tests (Section 2.5, ASTM D2265) confirmed that the palm grease formulations had dropping points between 50 and 55 °C (Table 4), without significant oil separation or flow, which was also confirmed through visual inspection and stable COF readings during tests. Furthermore, similar bio-greases with glycerol monostearate thickeners have been reported to withstand temperatures up to 70 °C without failure [61] since the thickener's high melting point (~80 °C) and hydrogen bonding with palm oil acids (e.g., palmitic/oleic acid) enhanced its thermal resilience.

Turning to the kinematic viscosity results, the grease sample based on palm oil demonstrates a viscosity value of 41 cSt at 40 °C, which is significantly lower than that of the commercial greases. However, the calculated viscosity index of palm grease is 209, which indicates more stability of palm oil properties at elevated temperatures in comparison with mineral oils used for lithium and sodium greases. Moreover, the base oil of the bio-grease sample reveals a significant value of the pour point. This behavior may be justified by the levels of saturated fatty acids, which constitute 52% of the structure [58,91]. The high saturation inhibits the grease's ability to flow at lower temperatures, thereby increasing its resistance to movement [53].

Table 4. Physical and chemical results of test grease samples.

Test Description	Test Standard	Lithium Grease	Sodium Grease	Palm Grease A
Unworked penetration test (dmm)	ASTM D217 [62,78]	250 (NLGI 3)	265 (NLGI 2)	185 (NLGI 4)
Static bleeding value 40 °C (oil mass %)	ASTM D1742 [81]	1–3	3–4	7.1
Dropping point (°C)	ASTM D2265 [79,80]	180–200	180–200	55–60
Kinematic viscosity at 40 °C (cSt)	ASTM D445 [77]	125	135	41
Kinematic viscosity at 100 °C (cSt)	ASTM D445 [77]	12	10	9
Viscosity Index (VI)		82	20	209
Pour point (°C)	ASTM D7346 [92]	−15	−5 to −10	9

3.3. *Estimated Lubrication Regime and Minimum Film Thickness*

It is crucial that the lubricant film generates an elastohydrodynamic lubrication condition between rolling elements and raceway surfaces to act as a capacitance and completely isolate the interfaces during the electric tests. Hence, each grease amount inside the rolling bearing was calculated. The developed oil film thickness during phase 1 of the electric test is a function of the bearing hertzian contact area, base oil viscosity, operating speed, and radial load.

To gain a deeper understanding of how the kinematic viscosity and consistency of each grease blend influence the developed film thickness, lubrication regime, and coefficient of friction (COF) of the test bearings, the minimum oil film thickness (H_{min}) is determined. The Hamrock–Dowson equation, recognized as the standard method for calculating the minimum film thickness in oil-lubricated rolling bearings, is utilized in this section for this purpose [93] as shown in (Equation (2)).

$$H_{min} = 3.63U^{0.68}.G^{0.49}.W^{-0.073}\left(1 - e^{-0.68k}\right) \quad (2)$$

where U is the speed parameter, G is the material parameter, and W is the loading parameter. Notably, the material parameter G is directly proportional to the pressure–viscosity coefficient (α) of the base oil in the grease, which defines how viscosity increases under pressure, critically affecting EHL film thickness [94]. For the studied greases, the used pressure–viscosity coefficient values are as follows: commercial lithium grease: $\alpha = 2.0 \times 10^{-8}$ Pa^{-1}, sodium grease: $\alpha = 0.5 \times 10^{-8}$ Pa^{-1}, and palm grease: 1.0×10^{-8} Pa^{-1} [95–97]. These values are obtained from published data on base oils and confirmed through comparison with reported film thickness calculations for similar lubricants [98]. To assess the lubrication regime of the grease under the selected speed and loading conditions, the λ parameter is computed (Equation (3)), representing the relationship between the minimum oil film thickness and the combined surface roughness of the interacting surfaces of the bearing raceways [99–101].

$$\lambda = \frac{h_{min}}{\sqrt{{R_{q1}}^2 + {R_{q2}}^2}} \quad (3)$$

Rq_1 represents the surface roughness value of the rolling element and Rq_2 is the corresponding surface roughness of the inner or outer raceway in contact. For the lubricated bearing to operate in the elastohydrodynamic lubricating (EHL) regime, λ values should be above 3. However, λ values between 1 and 3 indicate an operation of bearing in a mixed lubrication regime with partial film separation and moderate wear. On the other hand, boundary lubrication condition with direct metal contact takes place in case the λ value is less than 1, leading to high mechanical wear and a resistive state of current.

In the film thickness calculation, a temperature of 40 °C was assumed, which aligns with the design limits and operating conditions specified for the bearing–lubricant system under study. This temperature was selected based on expected thermal behavior during moderate-speed and moderate-load operation, as commonly reported in elastohydrodynamic lubrication (EHL) studies [102,103]. In the actual tests, bearing temperature was measured directly using a contact thermometer placed near the outer race of the bearing housing. The average temperature is recorded during each running test and was found to be 40 ± 4 °C, depending on lubricant type, applied voltage, and test duration. These values confirm that the assumed temperature for the film thickness model reasonably represents the real thermal environment during bearing operation [104].

There are several factors that control the dielectric strength and EDM voltage of lubricants in bearings subjected to parasitic currents such as oil film thickness, lubricant relative permittivity, and applied nano-additives. The minimum oil film thickness of the hertzian contact between rolling element and each raceway is one decisive factor in determining the capacitance and the dielectric strength of a typical lubricant type. According to capacitor theory, electric capacitance is inversely proportional to the film thickness. In a previous study [105], an experimental test setup was used to establish a more accurate simulation model that correlates the capacitance of a lubricant in deep-groove ball bearing and film thickness. In a more recent study, Bader et al. [106] experimentally confirmed that the oil film thickness is inversely proportional to the capacitance of the lubricant with increasing the operating speed. On the other hand, Maruyama et al. [107] developed electrical impedance methods to simultaneously measure oil film thickness and breakdown ratio in elastohydrodynamic (EHD) contacts of practical ball bearings. The breakdown ratio was found to increase as the oil film thickness decreases, indicating that thinner films are more prone to electrical breakdown under operational conditions.

The values of H_{min} and λ are calculated using Equations (2) and (3) for palm grease, lithium grease, and sodium grease; the values are presented in Table 5. All grease blends showed λ parameter values higher than 3, indicating that the test bearings operate predominantly in the EHL regime under the predefined operating speed and load. The differences between the results amongst the grease types are mainly attributed to the base oil viscosity since other parameters remained constant for all tested grease types. Test bearings lubricated with lithium grease and sodium grease showed calculated minimum oil film thickness (H_{min}) values of 7.28 μm and 8.52 μm at the inner raceway, respectively. In comparison, they are 8.60 μm and 9.65 μm at the outer raceway, respectively. The estimated values suggest that these greases have adequate viscosity values at the operating temperature, which are confirmed by the kinematic viscosity results in Table 4. Based on the previously mentioned works, it can be inferred that they possess the highest electric capacitance values amongst all grease blends and are expected to have a higher threshold for EDM voltage levels than palm grease samples. Palm grease A (without additives) exhibited lower (H_{min}) values (0.50 μm for the inner raceway and 0.59 μm for the outer raceway), corresponding to λ parameter higher than the threshold for EHL zone. Hence, the calculated ($H_{min)}$ values indicate that palm grease is expected to contribute to lowering the breakdown voltage value in comparison with lithium and sodium grease when subjected to electric field across the bearing rings.

Table 5. Calculated minimum film thickness H_{min} and λ parameter for each grease type.

Lubricant Blend	Operating Speed (rpm)	Bearing Radial Load (N)	Interacting Raceway	H_{min} (μm)	λ
Lithium grease	1400	100	Inner	1.03	7.28
			Outer	1.21	8.60
Sodium grease			Inner	1.20	8.52
			Outer	1.42	9.65
Palm grease A			Inner	0.50	3.53
			Outer	0.59	4.16

3.4. Tribological Results

According to the calculations of minimum oil film thickness, at the hertzian contact, all tested grease blends manage to operate the tested deep groove ball bearing in the EHL regime. This positively impacts reducing the friction coefficient during tribo-testing using the pin-on-disk tester. Figure 7 shows the calculated average COF values for three repeated tests for each grease type. It is observed that the COF of palm grease A (in its plain form) is lower than that of lithium grease and sodium grease by around 40% and 30%, respectively. For palm grease B, C, and D, the COF is decreased by 53%. The lowest COF value was recorded for palm grease D, reaching only 0.06.

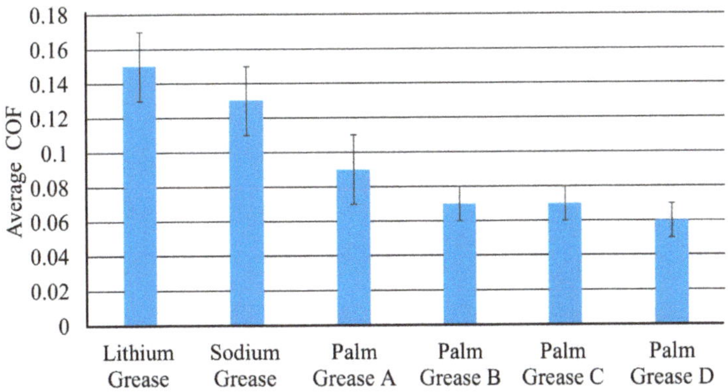

Figure 7. Average friction coefficient values for each test grease.

Both constituents of palm grease play a vital role in enhancing the friction behavior between the two mating surfaces of bearing components [61]. Glycerol as a thickener has polar molecules, namely the hydroxyl (–OH) groups, which create hydrogen bonds with oxides present on steel surfaces. This interaction creates a hydration layer, which reduces direct metal-to-metal contact and acts as a boundary lubricant [108,109]. Palm oil, on the other hand, is rich with long-chain fatty acids (palmitic acid, oleic acid, and linoleic acid). It adheres to surfaces primarily through free (unsaturated) fatty acids that interact with metal surfaces via chemisorption, forming a metal soap layer (metal–carboxylate complex). Although less interactive with metal surfaces, the non-polar hydrocarbon chains in palm oil from the saturated palmitic fatty acids still adhere to metal surfaces via Van der Waals forces, forming a thin, stable hydrophobic film that repels water and enhances lubrication. Hence, the multi-layered boundary films from the base oil and thickener reduce friction by preventing direct contact between sliding surfaces.

The addition of MWCNTs in palm grease blends (B, C, and D) further enhances the COF values by creating a protective tribofilm on contact surfaces, minimizing direct metal-to-metal interaction and reducing friction [110–113]. They can possibly act as nano-rollers and their rolling effect between frictional interfaces contributes to a lower COF by transforming sliding friction into a combination of rolling and sliding motion [114]. Moreover, MWCNTs absorb onto the metal surface via weak van der Waals interactions and/or π–π stacking with metallic surfaces containing carbonaceous layers to generate a thin lubricating layer, reducing direct contact between asperities [115].

3.5. Electrical Conductivity and EDM Voltage Analysis

The main focus of this work is to study the electric behavior of palm grease as an alternative to commercial greases in rolling bearings under electric fields. Additionally, the influence of MWCNT nano-additives at different concentrations of palm grease on the threshold and range of EDM voltage is being investigated. The bearing circuit current and the corresponding input DC voltage signal are continuously recorded for each test bearing, as described in Figure 3.

Figure 8 summarizes the obtained recorded electric results for each grease blend. In the case of lithium grease, the electric circuit remained in capacitive state at voltage amplitudes below 1.8 V. This phase is manifested by the absence of flowing electric current across the lubricant sides. When the input DC voltage value reached 1.8 V, repeated electric discharges started to appear in the monitored current signal, indicating the onset of EDM voltage occurrence. Increasing the input voltage beyond the threshold results in a corresponding rise in the severity of EDM current amplitudes. By reaching voltage levels of 7.5 V, both voltage and current signals manifest significant surges with rapid fluctuations, indicating that the grease in the hertzian contacts enters a resistive state to current flow. Sodium grease exhibited similar behavior to lithium grease, indicating comparable dielectric strength.

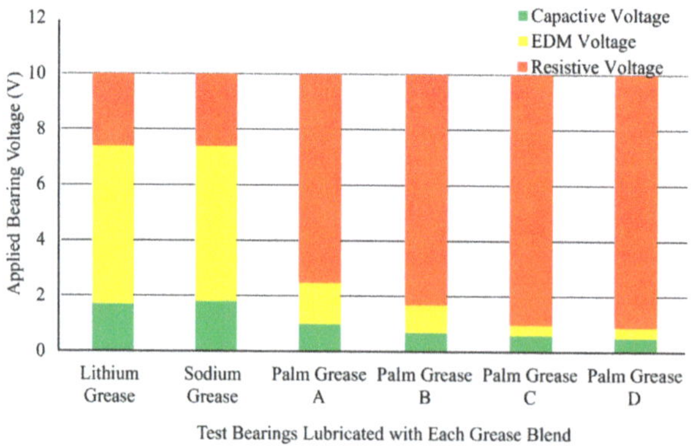

Figure 8. Measurement results of lubricant electric behavior under tested voltage steps between 1 and 10 V at 1400 rpm operating speed and a radial load of 100 N.

An example of the EDM voltage and current events for lithium grease are depicted in Figure 9 under 1.7 V amplitude. It is observed from the voltage signal that the lubricant undergoes rapid transient states of charging and discharging of current. During charging phase of the lubricant as a capacitance, the bearing voltage remains at an approximately constant value, then the amplitude suddenly drops to almost zero, marking the start of the discharging phase during which the insulating limit of the grease is reached, triggering the behavior of the lubricant to alternate between capacitive fluctuation from (EDM) current [116].

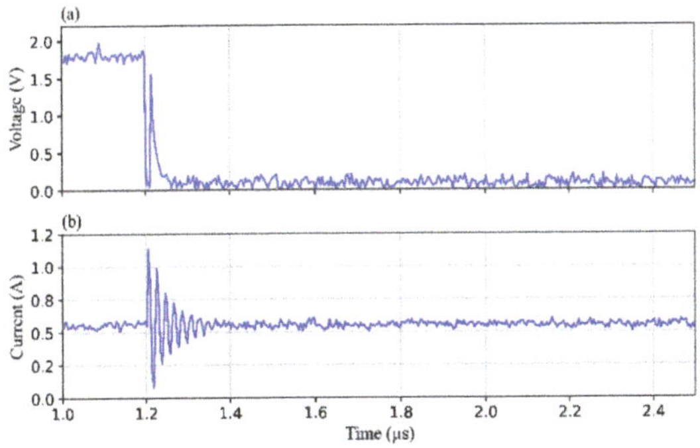

Figure 9. Electric charging and discharging event in rolling bearings lubricated with lithium grease: (**a**) EDM voltage and (**b**) current.

Palm grease A (in its plain form) showed a different electric capacitance response, with EDM voltage ranging from 1 V to 2.2 V only before entering the resistive state. This indicates that palm grease has lower breakdown voltage (BDV) than the two benchmark greases and operates more likely in the resistive state under considerable voltage values with limited possibilities for electrical discharge events to take place. The introduction of 1 wt.% MWCNTs into palm grease further reduced the BDV range to between 0.5 V and 0.9 V. This moderate decrease compared to plain palm grease suggests that MWCNT nanoparticles suspended in the grease and also adsorbed to the metallic surface provide channels for electrical current to flow at lower applied voltages, helping to limit the voltage fluctuations originating from EDM current charges. More enhancement was observed in the case of 2 wt.% and 3 wt.% MWCNTs where the voltage range reduced to only 0.6–1 V and 0.5–0.75 V, respectively. This substantial reduction in EDM voltage ranges suggests that MWCNTs increased the electric conductivity of the grease, preventing excessive EDM current effects, without affecting tribological performance as confirmed by Figure 7.

The presence of transient voltage spikes observed during testing is more accurately attributed to electrical discharge events (EDMs) occurring across the lubricant film, rather than indicating a complete transition to a resistive conduction state. These discharges arise when the applied voltage locally exceeds the BDV of the lubricant, i.e., resulting in momentary breakdowns of the insulating film. This phenomenon should be understood as statistical in nature, where the likelihood and frequency of discharge events increase progressively with rising electric field intensity. It does not represent a binary shift from capacitive to resistive behavior, but rather a growing probability of localized dielectric failure. The lubricant film may continue to exhibit predominantly capacitive characteristics while sporadically allowing discharges under high potential gradients. This interpretation is consistent with established research on electrostatic bearing degradation and discharge mechanisms in lubricated contacts under electric stress [31,117].

Palm oil has lower electrical conductivity than other vegetable oils with higher unsaturated fatty acid content. The presence of approximately 40–45% saturated palmitic acid (C16:0) reduces the number of polar functional groups, contributing to electrical conductivity [118]. A previous study was conducted by Slita et al. [119] to measure and compare the electrical resistivity of various vegetable oils. The results showed that palm oil had higher resistivity than other oils, indicating lower electrical conductivity. Yet, palm oil is still rich in monounsaturated oleic acid (C18:1), accounting for around 39–45%, polyunsaturated linoleic acid (C18:2) representing 10–11%, and minor traces of myristic acid (C14:0), lauric acid (C12:0), and linolenic acid (C18:3), as well as moisture [120]. Hence, the dielectric behavior (i.e., relative permittivity) of palm oil is significantly controlled by its fatty acid composition, particularly the balance between saturated and unsaturated fatty acids [121]. The presence of unsaturated fatty acids along with traces of moisture contributes to the increased molecular polarizability, which can enhance the dielectric constant in comparison with mineral oils existing in lithium and sodium greases. In a previous study by Rajab et al. [122], palm oil's dielectric constant and BDV was compared to mineral oil at different temperature ranges as an insulating liquid for transformer operations. It was found that the dielectric constant of palm oil is in the range of 3.0–3.2, while mineral oil is around 2.1–2.2 at room temperature. This is due to the higher degree of polarity in palm oil, which is defined by the high degree of unbalance in molecules' geometrical chemical structure. Furthermore, the BDV of palm oil at temperatures below 100 °C is lower than mineral oil as an insulating oil. It was justified that the relative amount of water concentration in palm oil is higher than the case of mineral oil, which enhanced the polarity and reduced BDV. Other researchers conformed with the previous results and proved that different vegetable oils exhibit higher dielectric constant values compared to mineral transformer oils [123]. Hossain et al. [121], on the other hand, compared palm oil along with other vegetable oils to mineral oils in terms of viscosity, BDV, and relative permittivity. Palm oil showed higher BDV than mineral oil, in contrast to the results of the current work. However, this is justified by the higher viscosity results of palm oil (4 times) than the mineral oil used.

The apparent contradiction of reduced friction and electrical breakdown, in the case of palm grease samples, despite the oil film being thinner than that of lithium and sodium grease, can be explained by the formation of a robust protective tribofilm on the bearing surface. This tribofilm arises from the polar fatty acid components of palm oil and is further enhanced by the presence of multi-walled carbon nanotubes (MWCNTs) in the experimental blends. The tribofilm effectively reduces direct metal-to-metal contact, acting as a physical barrier that lowers friction and mitigates electrical discharge machining (EDM) events by providing conductive pathways that dissipate charge and prevent severe electrical breakdown. Moreover, while transient local mixed or boundary lubrication conditions may arise during electrical discharge events, the tribofilm buffers these occurrences, maintaining lubrication efficiency and protecting against wear. Thus, the lubrication regime with palm grease should be characterized as primarily EHL combined with dynamic mitigation of localized boundary interactions through tribofilm formation, resulting in superior triboelectric performance compared to traditional lithium and sodium greases.

3.6. Vibration Analysis Results

Following the electric tests, the damage level in each lubricated test bearing is assessed in this section by operating the bearing under the same operating conditions in the test rig while switching off the electric circuit. Vibrations are measured in the time domain and frequency domain to evaluate the induced faults within the bearing elements using a radially mounted accelerometer to the bearing hub [124,125]. The electric sparks generated during EDM discharges can produce extremely high temperatures, which typically range up to several thousand degrees Celsius [117]. These intense temperatures are sufficient to locally melt and vaporize bearing surfaces, leading to surface localized faults such as spalling and micro-pitting. Frequent melting on localized areas on the surface may cause material quenching in the presence of base oil and gradually generate frosting or distributed roughing of the surface.

Each time the rolling elements pass over a localized fault, it induces recurring temporary impulses in the measured vibration acceleration signal at one of the bearing characteristic frequencies in the low-to-medium frequency range [126,127]. By this time, thousands of micro-pits are generated, resulting in distributed faults along the raceway of rough-surface-inducing harmonics and sidebands to the bearing fault frequency signal and worsening the vibration levels. An example of the time waveform and frequency spectrum for an already electrocuted test bearing with palm grease is shown in Figure 10. The time domain shows the effect of the damage on the raceways in the form of transient spikes where the time between each consecutive impulses represents the frequency of the bearing fault. The frequency domain shows peaks at 1x of the operating speed with its harmonics resembling the excessive clearance in the bearing due to wear. The bearing fault frequency and its harmonics are raised by a hump of energy, indicating broadband noise and late stage of bearing faults before failure.

(a)
(b)

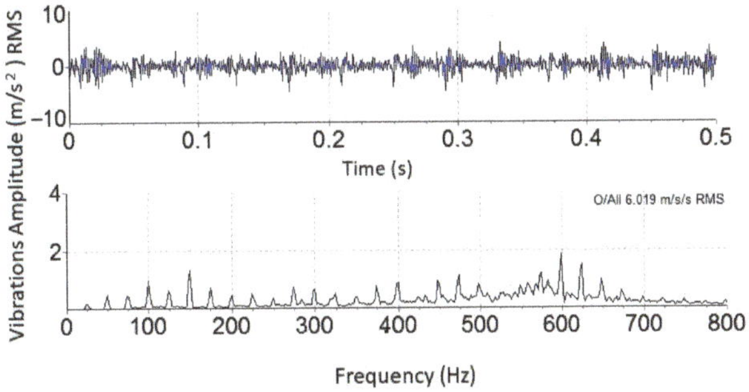

Figure 10. Vibrations signal analysis of bearings lubricated with palm grease A in (**a**) Time waveform and (**b**) frequency spectrum.

Figure 11 shows the summary of the overall amplitudes of vibration spectra in RMS that are calculated for each bearing. It is observed that sodium grease exhibits the highest vibration levels among the tested lubricants, indicating severe electric-based damage on the bearing raceways and rolling element during operation. Lithium grease, on the other hand, significantly improves vibrations, with 23% and 25% drop in overall levels in the vertical and horizontal directions, respectively, when compared with sodium grease. This means less induced localized and distributed faults on the bearing raceways when using lithium grease. It also indicates that lithium grease provides better protection against electric sparks while providing better tribological behavior.

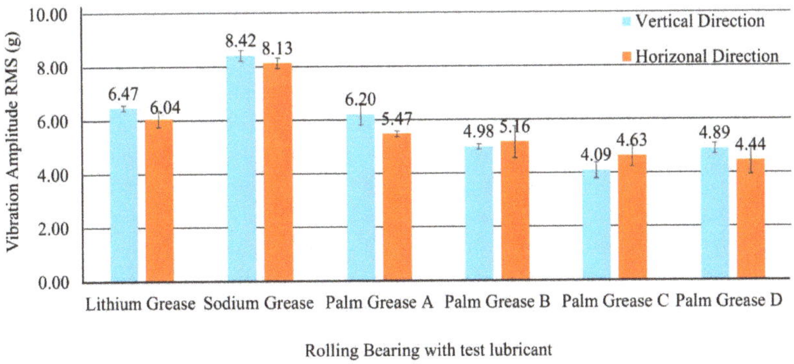

Figure 11. Vibration levels measured in vertical and horizontal directions for each test bearing.

A similar improvement was detected for test bearings lubricated with palm grease. In its plain form, palm grease reduced the vibrations amplitudes by 26.5% in the vertical direction and 33% in the horizontal direction, compared with sodium grease. When MWCNTs are introduced into palm grease, the vibration levels further decrease, highlighting the impact of nanoparticles in mitigating the damage-induced vibrations. For instance, bearings lubricated with palm grease B (blended with 1 wt.% MWCNTs) reduced the vibration levels (in the vertical direction) by 23% and 41% compared to lithium grease and sodium grease, respectively.

The most significant improvement is observed with palm grease C and D, which achieve the lowest vibration levels among all tested greases. These formulations reduced vertical and horizontal vibrations by 29% and 33% compared to lithium grease, respectively. The presence of MWCNTs in palm grease contributed to lower EDM voltage threshold, leading to expected less intense sparks between rolling elements and raceways. While the EDM discharges intense heat, which tends to degrade the grease components, MWCNTs on the mating surfaces of the already electrocuted bearing raceways have the potential of creating mechanical interlocking and fill-in effects. They penetrate surface asperities, smoothing rough surfaces and filling micro-pits. This mitigates the resultant mechanical vibrations from localized faults (such as micro-pits and spalls) and distributed faults (frosting rough texture) on the surface.

Figure 12 demonstrates a section of the test bearing lubricated with lithium grease after the electric test run. The raceway of the inner ring reveals the generation of gray frosting in the middle region. In contrast, the lithium grease on the raceway, the cage, and rolling element surfaces suffered a discoloration from bright to dark yellow and gray. Test bearings lubricated with other grease blends experienced a similar damage pattern.

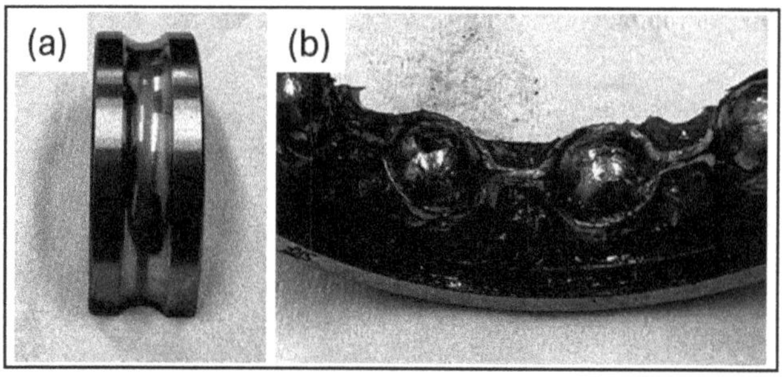

Figure 12. A section in the damaged ball bearing after electric test showing evidence of (**a**) frosting on the inner raceway and (**b**) grease discoloration between the cage and balls.

3.7. Surface Damage Analysis

For more profound understanding of vibrations and tribo-electric results, SEM examination is conducted on the worn surfaces of sectioned raceways of each test bearing subjected to the electric stress at a resolution of 10 μm and 1000× magnification, as shown in Figure 13. The images depict damaged raceway surface texture and faults to provide more information on the impact of each grease blend on the bearing tribo-electric performance when subjected to the parasitic currents. Further analysis is conducted using energy dispersive X-ray analysis (EDX) using (JEOL JSM-IT200, Akishima, Tokyo, Japan) to discover evidence of the deposited chemical elements on the worn surfaces of the inner and outer raceways of the bearing and to explain the damage mechanism and severity. In Figure 14a–f, significant concentrations of Fe, Cr, and Mn are detected on the surface, confirming the presence of steel alloy constituents of the bearing raceway material.

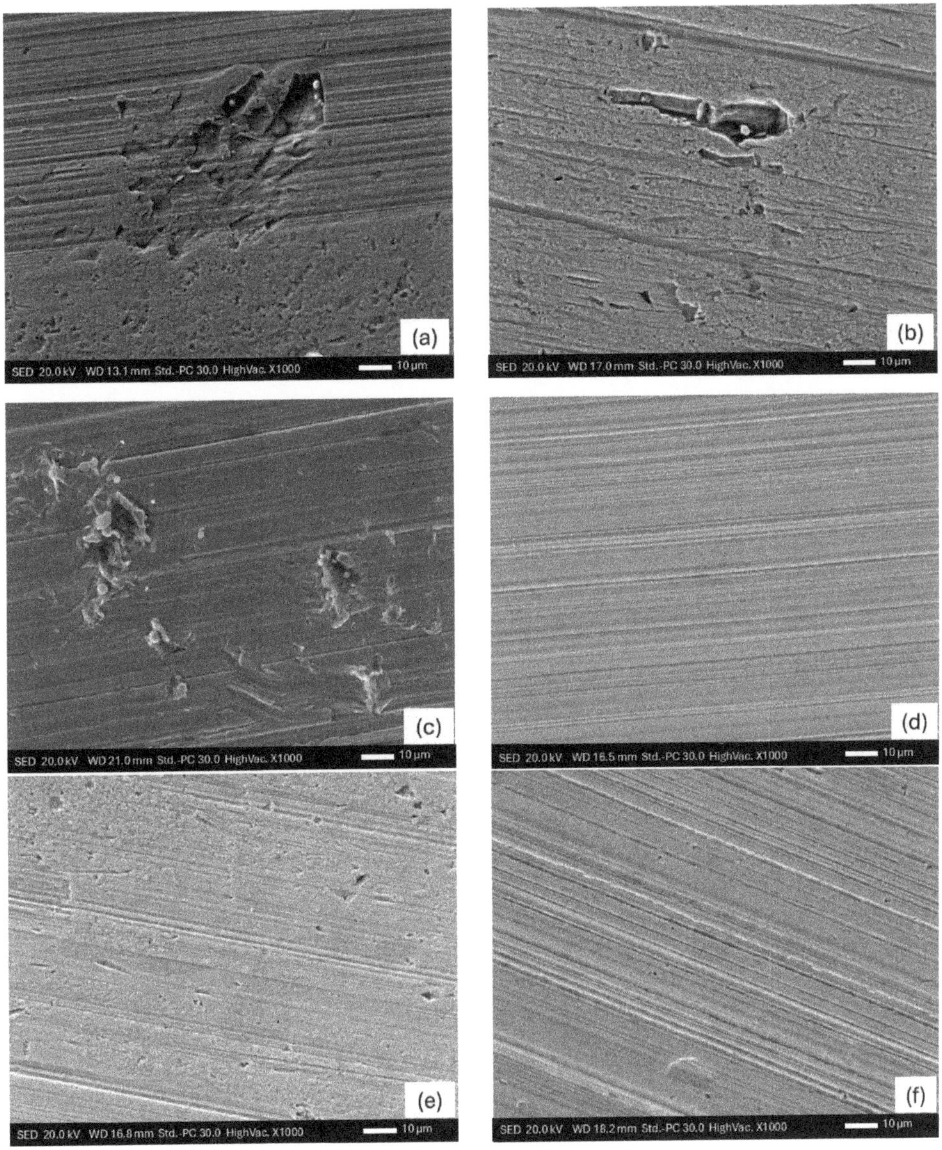

Figure 13. SEM images at magnification 1000× of the bearing worn surface for test samples: (**a**) lithium grease; (**b**) sodium grease; (**c**) palm grease without additives; (**d**) palm grease with 1 wt.% MWCNTs; (**e**) palm grease with 2 wt.% MWCNTs; (**f**) palm grease with 3 wt.% MWCNTs.

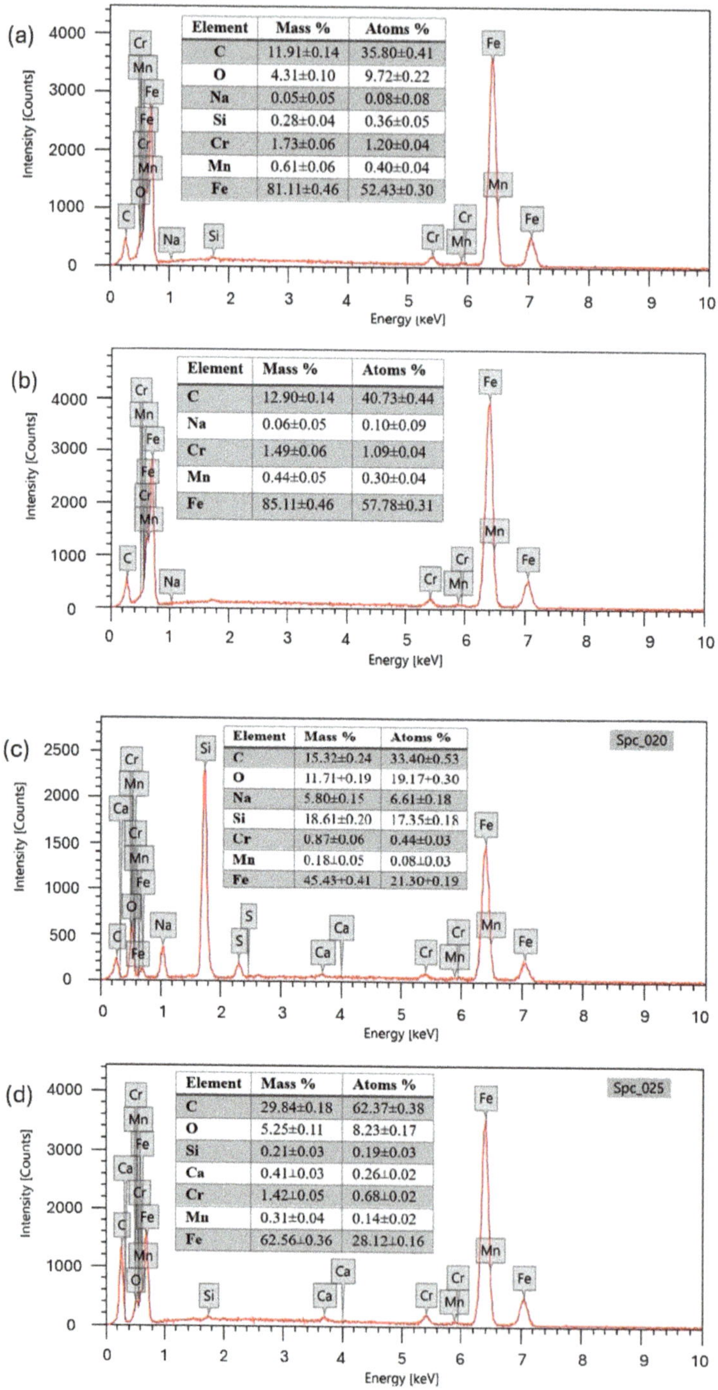

Element	Mass %	Atoms %
C	11.91±0.14	35.80±0.41
O	4.31±0.10	9.72±0.22
Na	0.05±0.05	0.08±0.08
Si	0.28±0.04	0.36±0.05
Cr	1.73±0.06	1.20±0.04
Mn	0.61±0.06	0.40±0.04
Fe	81.11±0.46	52.43±0.30

Element	Mass %	Atoms %
C	12.90±0.14	40.73±0.44
Na	0.06±0.05	0.10±0.09
Cr	1.49±0.06	1.09±0.04
Mn	0.44±0.05	0.30±0.04
Fe	85.11±0.46	57.78±0.31

Element	Mass %	Atoms %
C	15.32±0.24	33.40±0.53
O	11.71±0.19	19.17±0.30
Na	5.80±0.15	6.61±0.18
Si	18.61±0.20	17.35±0.18
Cr	0.87±0.06	0.44±0.03
Mn	0.18±0.05	0.08±0.03
Fe	45.43±0.41	21.30±0.19

Element	Mass %	Atoms %
C	29.84±0.18	62.37±0.38
O	5.25±0.11	8.23±0.17
Si	0.21±0.03	0.19±0.03
Ca	0.41±0.03	0.26±0.02
Cr	1.42±0.05	0.68±0.02
Mn	0.31±0.04	0.14±0.02
Fe	62.56±0.36	28.12±0.16

Figure 14. *Cont.*

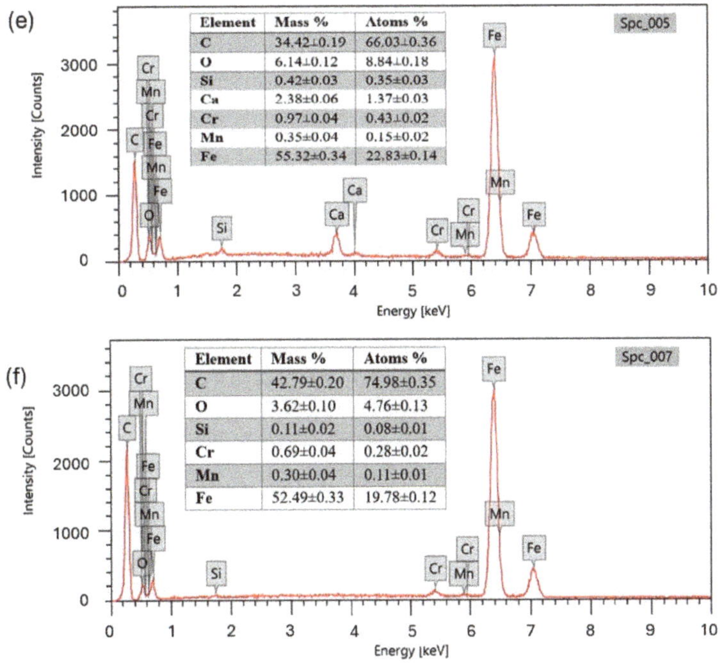

Figure 14. EDX spectra of the bearing raceway worn surface for test samples: (**a**) lithium grease; (**b**) sodium grease; (**c**) palm grease without additives; (**d**) palm grease with 1 wt.% MWCNTs; (**e**) palm grease with 2 wt.% MWCNTs; (**f**) palm grease with 3 wt.% MWCNTs.

The formation of such tribofilm was confirmed via SEM-EDX examination of the bearing raceway, which revealed a smoother worn surface and the presence of a uniform film-like layer on the contact track for MWCNT-enhanced grease samples compared to unmodified grease. EDX analysis of the same regions further supported this finding by showing elevated carbon signals and reduced metal peak intensities, indicating that the surface was covered with a carbonaceous layer attributed to MWCNTs. Tribofilm involving carbonaceous layers, metal oxides, or organometallic compounds have been reported when MWCNTs were used under electrical stress.

Figure 13a shows the wear scar surface of the inner race of the ball bearing sample lubricated with lithium grease. Widely distributed micro-pits with frosting of a dull appearance and rough morphology can be observed on the examined worn surface. The micro-pits and large craters indicate severe surface damage induced by electric discharge sparks during the test duration, leading to localized melting and material removal. The repeated discharges cause surface alterations, leading to the characteristic frosted appearance. Based on the calculated film thickness and lambda values for lithium grease, the developed tribo-layer during running tests was expected to sufficiently shield the contact surfaces, leading to adequate protection against mechanical rubbing between the rolling elements and raceways. Mechanical damage in the form of deep furrow marks also occurs due to abrasive wear in the rolling direction. This is justified by the fact that micro-arcing, during EDM current phase, creates cratering and pitting, releasing hardened metallic debris into the lubricant grease and encouraging three-body wear mechanisms. In addition, the rapidly repetitive EDM currents degrade and carbonize the grease blend within the bearing, producing byproduct sludge and abrasive residues. The loss of lubricant film integrity allows contaminants to embed into surfaces, accelerating the three-body abrasion. The previously mentioned faults on the bearing raceway were manifested by the significant vibration levels (RMS) measured on the test bearing in the radial direction. According to Reddy et al. [128], commercial lithium grease possesses low electric conductivity even with lithium ions within the thickener structure.

This increased the electric impedance of the sides of the lubricant film acting as a capacitor. Gradually, voltage amplitudes will build up across the capacitor, leading to significant discharges of intensive heat that eventually evaporate the base oil and create micro-pits across the raceways by electric erosion [129]. In Figure 14a, the EDX analysis of commercial lithium grease reveals moderate carbon content (11.91%) and oxygen content (4.31%) along with traces of Na (0.05%), indicating a remarkable degradation of grease structure under electric stress into its basic elements. This conforms with the SEM findings, which show the rapid evaporation of lubricant film, leaving the raceway surfaces prone to both mechanically abrasive wear and electrical-induced spalling and frosting.

Similar worn surface results are found in the case of sodium grease-lubricated bearings as shown in Figure 13b. Rolling tracks beside narrow and deep grooves can be observed on the examined worn surface [130] caused by secondary damage mechanism from micro-pitting of electrocuted surfaces. Furthermore, obvious frosting texture is detected due to cyclic melting of the surface during the EDM voltage application. Grease components (base oil and soap) decomposed, leaving behind carbonaceous residues and sodium on the raceway surface accelerating the mechanical abrasive wear mechanism.

The elemental analysis of the worn surfaces confirms this. The EDX result revealed a carbon content of 12.9% with extremely low O and Na contents, as seen in Figure 14b. In Figure 13c, the raceway worn surface lubricated with palm grease is mainly shallow with abrasive friction tracks and irregular embossments on the surface from localized melting and spalling.

The fatty acids in palm oil contributed to the reduction in mechanical damage on the raceway surface while also improving the coefficient of friction (COF). This effect can be attributed to their long molecular chains and polar functional groups, which promote strong interactions with the rubbing contact surfaces [62]. In addition, frosting and micro-pits appeared in the case of palm grease as they showed a similar EDM voltage threshold to sodium grease. Figure 14c illustrates the EDX results for the case of palm grease. EDX measurements proved 15.32% carbon content and the highest oxygen content (19.17%). Since the palmitic and stearic acids are long-chain fatty acids bonded to a glycerol backbone, they possess higher molecular weight organic compounds with a more significant proportion of carbon, hydrogen, and oxygen. This is reflected in the increase in the carbon and oxygen content deposited on the surface compared to lithium and sodium grease. Unexpected presence of Si content appears in the EDX spectrum for palm grease samples. Since palm grease in principle does not have silicon-based additives, the detected Si is thus not from intrinsic grease components. The most plausible source of Si could be debris trapped and deposited on the bearing surface due to the thinner lubrication film formed by the palm grease.

For palm grease with MWCNT additives (Figure 13d–f), the worn surface images show that the furrows become superficial and mild with a smoother topography. From the tribological perspective, it is established that the self-lubricating characteristic of the carbonaceous material greatly reduces the COF compared to lithium, sodium, and palm grease, which is reflected on the enhanced worn surface results. Also, the surfaces lack obvious micro-pits or frosting, suggesting a significant impact of MWCNTs in reducing the surface-induced electric faults. This can be justified by the electric stress results where palm grease with MWCNTs exhibited lower EDM voltage margins than other grease types. The palm grease formulation used in this work does not contain any calcium-based thickeners.

Therefore, the calcium detected in the EDX analysis of bearing surfaces lubricated with palm grease containing 2 wt.% MWCNTs (Figure 14e) is likely due to environmental contamination or wear debris from bearing components. The presence of MWCNTs may enable the adsorption or localization of Ca ions onto the nanotube surfaces, enhancing their detectability in EDX.

Previous works such as Jonjo et al. [46] proved that the addition of MWCNTs to lithium grease enhanced its electric conductivity and mitigated the surface damage during EDM discharge events. The formation of physical adsorption tribofilm from conductive nano-additives plays a crucial role in facilitating conductive paths for current through the grease structure [131]. The enhancement in electric conductivity reduces the charging time across the lubricant bearing interfaces, resulting in less intense spark to discharge and hence lower damage to the surface of bearings. EDX results (Figure 14d–f) revealed carbon content higher than 60% on the worn surfaces of tested bearings, proving the presence of large residuals of MWCNT layer, confirming the mechanical wear resistance, and less electrical fault features. Palm grease with 2% MWCNTs appears to provide the best balance between lubrication efficiency and oxidative stability among the tested samples.

The service life of the newly developed palm grease formulations containing multi-walled carbon nanotubes (MWCNTs) is dependent on their thermal, oxidative, mechanical, and electrical stability. In this study, grease samples were tested under controlled conditions, including a radial load of 100 N, an average operating temperature of 40 °C, and incremental DC voltages up to 10 V. These conditions simulated moderate stress levels over continuous 6 h test durations, during which no signs of base oil separation, hardening, or degradation were observed. This suggests that the grease maintains functional stability during medium-term operation under typical loading and temperature conditions [132].

The enhanced electrical conductivity observed in MWCNT-doped palm greases can be attributed to the formation of a percolation network, which facilitates electron transport across the lubricant film. This conductivity improvement contributes to lowering the electrical discharge machining (EDM) threshold by enhancing charge dissipation. The network formed by well-dispersed nanotubes remains stable as long as there is no significant mechanical shear degradation or thermal oxidation. Previous studies have shown that carbon-based nanofillers, when adequately dispersed, can sustain enhanced conductivity and electrostatic discharge mitigation capabilities over extended use [132,133].

Nonetheless, over long-term operation, the electrical performance of the grease may decline due to potential agglomeration of MWCNTs, thermal aging of the palm oil base, and mechanical breakdown of the percolated structure. Such effects can reduce conductivity and alter the tribo-electrical behavior. To quantify this decline, standardized long-duration testing such as ASTM D3336 (Grease Life in Ball Bearings at Elevated Temperatures) is recommended, along with extended cyclic electrical stress testing to observe discharge behavior over time [134]. These protocols would provide a more accurate assessment of the grease's operational life and its electro-functional stability under realistic EV conditions.

4. Conclusions

A novel palm grease, enhanced with MWCNTs, is synthesized and tested against commercial greases in tribo-electric environment. Through rigorous chemical–physical, tribological, electrical analyses, vibrational, and SEM-EDX analyses, this study reveals the following key conclusions:

1. The palm-based bio-grease demonstrated superior consistency (NLGI 4) and thermal stability compared to commercial lithium and sodium greases, with a high viscosity index (209) ensuring performance across temperatures.

2. Palm grease reduced the coefficient of friction (COF) by 40% versus lithium grease and 30% versus sodium grease. Incorporating multi-walled carbon nanotubes (MWCNTs) further lowered COF to 0.06 (3 wt.%), reducing friction by up to 60 %. The results indicate better physical and chemical adsorption of palm grease to the surface, providing a protective layer.
3. Benchmark greases resulted in higher EDM voltage thresholds and ranges, with lithium grease entering the EDM phase at 1.7 V and sodium grease at 1.2 V. Exposed bearing raceways to these frequent current discharges resulted in severe localized and distributed surface damage along with high vibration levels in subsequent bearing runs.
4. Palm grease exhibited a lower EDM voltage range (1.0–2.2 V), reducing harmful bearing currents. MWCNT additives (2–3 wt.%) further decreased EDM thresholds (0.5–0.75 V), mitigating surface damage from electric discharges as evident by subsequent vibration analysis and SEM-EDX examination of damaged surface. Palm grease with 2 wt.% MWCNTs achieves optimal vibration damping, reducing vertical and horizontal vibrations by 28.41 % and 32.37 %, respectively, outperforming commercial greases.

In summary, the bio-derived palm grease is found to mitigate harmful electric erosion due to EDM voltage with no compromise in tribological performance. The palm grease with MWCNTs as nano-additives is believed to form a protective tribofilm, reducing direct metal-to-metal contact and improving lubrication efficiency. Its comparable performance to lithium grease makes it a potential lubricant for challenging applications such as electric mobility.

Future work is devoted to study the rheological and thermal behavior of palm grease under different operating conditions including a clear identification of the grease operating temperature limits and the enhancement of the oxidation stability. It is recognized that AC and PWM voltages can induce more severe discharge events, including characteristic fluting damage, which may not be fully replicated under DC conditions. Future work will extend this investigation to include AC and PWM excitation to more closely simulate real-world EV-operating environments Furthermore, a comprehensive investigation of the durability of bearings lubricated with palm-based nano-lubricants is considered for further study to accurately assess how each grease influences bearing the attainable fatigue life factors under electrical and mechanical stresses.

Author Contributions: Conceptualization, M.G.A.N., M.E.-H., M.M.Y. and A.A.A.; Methodology, A.A.A., M.M.Y. and M.G.A.N.; Samples' Preparation, A.A.A., M.M.Y. and M.G.A.N.; Formal Analysis, M.G.A.N., T.F.M. and A.A.A.; Investigation, A.A.A., M.G.A.N., M.M.Y. and F.P.; Resources, T.F.M. and M.G.A.N.; Data Curation, M.G.A.N. and A.A.A.; Writing—Original Draft Preparation, A.A.A. and M.G.A.N.; Writing—Review and Editing, M.E.-H., M.M.Y., A.A.A., F.P. and M.G.A.N.; Visualization, M.G.A.N. All authors have read and agreed to the published version of the manuscript.

Funding: This research received no external funding.

Data Availability Statement: The original contributions presented in this study are included in the article. Further inquiries can be directed to the corresponding authors.

Conflicts of Interest: The authors declare no conflicts of interest.

References

1. SEI; Climate Analytics; E3G; IISD; UNEP. Phasing Down or Phasing Up? Top Fossil Fuel Producers Plan Even More Extraction Despite Climate Promises: Production Gap Report 2023. Available online: https://www.sei.org/publications/production-gap-report-2023/ (accessed on 10 June 2025).
2. Perera, F. Pollution from Fossil-Fuel Combustion Is the Leading Environmental Threat to Global Pediatric Health and Equity: Solutions Exist. *Int. J. Environ. Res. Public Health* **2017**, *15*, 16. [CrossRef] [PubMed]
3. IEA. *CO_2 Emissions in 2023*; IEA: Paris, France, 2024. Available online: https://www.iea.org/reports/co2-emissions-in-2023 (accessed on 10 June 2025).
4. U.S. Environmental Protection Agency (EPA). Carbon Pollution from Transportation. Transportation, Air Pollution, and Climate Change Program. Last Updated 25 April 2025. Available online: https://www.epa.gov/transportation-air-pollution-and-climate-change/carbon-pollution-transportation (accessed on 20 June 2025).
5. Jose, P.S.; Natarajan, A.; Karthikeyan, S.; Bogaraj, T. Environmental and Social Impact of Electric Vehicles. In *Advanced Technologies in Electric Vehicles*; Gali, V., Canha, L.N., Resener, M., Ferraz, B., Varaprasad, M.V.G., Eds.; Academic Press: Cambridge, MA, USA, 2024; pp. 107–125. [CrossRef]
6. IEA. *Global EV Outlook 2024*; IEA: Paris, France, 2024. Available online: https://www.iea.org/reports/global-ev-outlook-2024 (accessed on 20 June 2025).
7. Li, S.; Tong, L.; Xing, J.; Zhou, Y. The Market for Electric Vehicles: Indirect Network Effects and Policy Design. *J. Assoc. Environ. Resour. Econ.* **2017**, *4*, 89–133. [CrossRef]
8. European Environment Agency. *Electric Vehicles from Life Cycle and Circular Economy Perspectives*; TERM Report 2018; Publications Office of the European Union: Luxembourg, 2018; ISBN 978-92-9213-985-8. [CrossRef]
9. Boztas, G.; Yildirim, M.; Aydogmus, O. Design and Analysis of Multi-Phase BLDC Motors for Electric Vehicles. *Eng. Technol. Appl. Sci. Res.* **2018**, *8*, 2646–2650. [CrossRef]
10. Yildirim, M.; Polat, M.; Kürüm, H. A Survey on Comparison of Electric Motor Types and Drives Used for Electric Vehicles. In Proceedings of the 2014 16th International Power Electronics and Motion Control Conference and Exposition (EPE-PEMC 2014), Antalya, Turkey, 21–24 September 2014; pp. 218–223. [CrossRef]
11. Chun, Y.-D.; Park, B.-G.; Kim, D.-J.; Choi, J.-H.; Han, P.-W.; Um, S. Development and Performance Investigation on a 60kW Induction Motor for EV Propulsion. *J. Electr. Eng. Technol.* **2016**, *11*, 609–617. [CrossRef]
12. Nam, K.H. *AC Motor Control and Electrical Vehicle Applications*, 2nd ed.; CRC Press: Boca Raton, FL, USA, 2018; 574p. [CrossRef]
13. Qureshi, M.S.; Kerai, A.A.; Fatima, S.A.; Shah, S.J.; Khan, K.A.; Ali, A.; Maheshwari, L.; Makda, I.; Usman, A. Effects of Parasitic Elements in High Frequency GaN-based DC–DC Converters for Electric Vehicle Applications. In Proceedings of the 2023 25th International Multitopic Conference (INMIC), Lahore, Pakistan, 4–5 November 2023; pp. 1–6. [CrossRef]
14. Landa, T.V.; Moore, A. Analysis of Parasitic Components in the PWM Inverter in an Electric Vehicle. *Int. J. Eng. Res. Technol.* **2013**, *2*, IJERTV2IS90671.
15. Busse, D.; Erdman, J.; Kerkman, R.J.; Schlegel, D.; Skibinski, G. Bearing Currents and Their Relationship to PWM Drives. *IEEE Trans. Power Electron.* **1997**, *12*, 243–252. [CrossRef]
16. Mütze, A. *Bearing Currents in Inverter-Fed AC Motors*, 1st ed.; Shaker-Verlag GmbH: Aachen, Germany, 2004; ISBN 978-3-8322-2528-5. Available online: https://tugraz.elsevierpure.com/en/publications/bearing-currents-in-inverter-fed-ac-motors (accessed on 20 June 2025).
17. Schneider, V.; Behrendt, C.; Höltje, P.; Cornel, D.; Becker-Dombrowsky, F.M.; Puchtler, S.; Gutiérrez Guzmán, F.; Ponick, B.; Jacobs, G.; Kirchner, E. Electrical Bearing Damage, A Problem in the Nano- and Macro-Range. *Lubricants* **2022**, *10*, 194. [CrossRef]
18. Loos, J.; Bergmann, I.; Goss, M. Influence of High Electrical Currents on WEC Formation in Rolling Bearings. *Tribol. Trans.* **2021**, *64*, 708–720. [CrossRef]
19. Mütze, A.; Binder, A. Techniques for Measurement of Parameters Related to Inverter-Induced Bearing Currents. *IEEE Trans. Ind. Appl.* **2007**, *43*, 1274–1283. [CrossRef]
20. Wang, F. Motor Shaft Voltages and Bearing Currents and Their Reduction in Multilevel Medium-Voltage PWM Voltage-Source-Inverter Drive Applications. *IEEE Trans. Ind. Appl.* **2000**, *36*, 1336–1341. [CrossRef]
21. Maruyama, T.; Nakano, K. In Situ Quantification of Oil Film Formation and Breakdown in EHD Contacts. *Tribol. Trans.* **2018**, *61*, 1057–1066. [CrossRef]
22. Binder, A.; Muetze, A. Scaling Effects of Inverter-Induced Bearing Currents in AC Machines. *IEEE Trans. Ind. Appl.* **2008**, *44*, 769–776. [CrossRef]
23. Collin, R.; Yokochi, A.; von Jouanne, A. EDM Damage Assessment and Lifetime Prediction of Motor Bearings Driven by PWM Inverters. In Proceedings of the 2022 IEEE Energy Conversion Congress and Exposition (ECCE), Detroit, MI, USA, 9–13 October 2022; pp. 1–6. [CrossRef]
24. Xu, Y.; Liang, Y.; Yuan, X.; Wu, X.; Li, Y. Experimental Assessment of High Frequency Bearing Currents in an Induction Motor Driven by a SiC Inverter. *IEEE Access* **2021**, *9*, 40540–40549. [CrossRef]

25. Im, J.-H.; Kang, J.-K.; Lee, Y.-K.; Hur, J. Shaft Voltage Elimination Method to Reduce Bearing Faults in Dual Three-Phase Motor. *IEEE Access* **2022**, *10*, 81042–81053. [CrossRef]
26. Muetze, A.; Oh, H.W. Application of Static Charge Dissipation to Mitigate Electric Discharge Bearing Currents. *IEEE Trans. Ind. Appl.* **2008**, *44*, 135–143. [CrossRef]
27. Ma, J.; Xue, Y.; Han, Q.; Li, X.; Yu, C. Motor Bearing Damage Induced by Bearing Current: A Review. *Machines* **2022**, *10*, 1167. [CrossRef]
28. Muetze, A.; Binder, A. Calculation of Influence of Insulated Bearings and Insulated Inner Bearing Seats on Circulating Bearing Currents in Machines of Inverter-Based Drive Systems. *IEEE Trans. Ind. Appl.* **2006**, *42*, 965–972. [CrossRef]
29. Hoffe, S.J.; Meyer, A.S.; Cronje, W.A. Determination of Shaft Position from Shaft Voltage on a Synchronous Generator. In Proceedings of the 2005 5th IEEE International Symposium on Diagnostics for Electric Machines, Power Electronics and Drives (DEMPED), Vienna, Austria, 7–9 September 2005; pp. 1–4. [CrossRef]
30. CW. *E-Mobility Bearings: Customized Ball Bearings for E-Mobility Drives*; CW Bearing GmbH: Hockenheim, Germany. Available online: https://www.izb-online.com/fileadmin/_migrated/exhibitor_database/upload/257/CW-E-Mobility-Brochure-EU.pdf (accessed on 20 June 2025).
31. He, F.; Xie, G.; Luo, J. Electrical Bearing Failures in Electric Vehicles. *Friction* **2020**, *8*, 4–28. [CrossRef]
32. Xie, G. Lubrication Under Charged Conditions. *Tribol. Int.* **2015**, *84*, 22–35. [CrossRef]
33. Xie, G.; Si, L.; Guo, D.; Liu, S.; Luo, J. Interface Characteristics of Thin Liquid Films in a Charged Lubricated Contact. *Surf. Interface Anal.* **2015**, *47*, 315–324. [CrossRef]
34. Fan, X.; Xia, Y.; Wang, L. Tribological Properties of Conductive Lubricating Greases. *Friction* **2014**, *2*, 343–353. [CrossRef]
35. Freschi, M.; Paniz, A.; Cerqueni, E.; Colella, G.; Dotelli, G. The Twelve Principles of Green Tribology: Studies, Research, and Case Studies—A Brief Anthology. *Lubricants* **2022**, *10*, 129. [CrossRef]
36. Bustami, B.; Rahman, M.M.; Shazida, M.J.; Islam, M.; Rohan, M.H.; Hossain, S.; Nur, A.S.M.; Younes, H. Recent Progress in Electrically Conductive and Thermally Conductive Lubricants: A Critical Review. *Lubricants* **2023**, *11*, 331. [CrossRef]
37. Walther, H.C.; Holub, R.A. Lubrication of Electric Motors as Defined by IEEE Standard 841-2009, Shortcomings and Potential Improvement Opportunities. In Proceedings of the 2014 IEEE Petroleum and Chemical Industry Technical Conference (PCIC), San Francisco, CA, USA, 8–10 September 2014; pp. 91–98. [CrossRef]
38. Mustafa, W.A.A.; Dassenoy, F.; Sarno, M.; Senatore, A. A review on potentials and challenges of nanolubricants as promising lubricants for electric vehicles. *Lubr. Sci.* **2022**, *34*, 1–29. [CrossRef]
39. Parenago, O.P.; Lyadov, A.S.; Maksimov, A.L. Development of Lubricant Formulations for Modern Electric Vehicles. *Russ. J. Appl. Chem.* **2022**, *95*, 765–774. [CrossRef]
40. Pape, F. Investigation of Graphene Platelet-Based Dry Lubricating Film Formation in Tribological Contacts. *Coatings* **2024**, *14*, 360. [CrossRef]
41. Duan, L.; Li, J.; Duan, H. Nanomaterials for Lubricating Oil Application: A Review. *Friction* **2023**, *11*, 647–684. [CrossRef]
42. Nassef, M.G.A.; Soliman, M.; Nassef, B.G.; Daha, M.A.; Nassef, G.A. Impact of Graphene Nano-Additives to Lithium Grease on the Dynamic and Tribological Behavior of Rolling Bearings. *Lubricants* **2022**, *10*, 29. [CrossRef]
43. Pape, F.; Poll, G. Investigations on Graphene Platelets as Dry Lubricant and as Grease Additive for Sliding Contacts and Rolling Bearing Application. *Lubricants* **2020**, *8*, 3. [CrossRef]
44. Liu, J.; Bai, Q.; Li, X.; Zhang, L.; Pape, F.; Guo, F.; Poll, G. Evolution Processes of Electrical Discharge in EHD Contact Lubricated with Conductive Grease. *Tribol. Int.* **2025**, *209*, 110725. [CrossRef]
45. Zhang, P.; Gong, J.; Jiang, Y.; Long, Y.; Lei, W.; Gao, X.; Guo, D. Application of Silver Nanoparticles in Parasite Treatment. *Pharmaceutics* **2023**, *15*, 1783. [CrossRef]
46. Jonjo, E.R.; Ali, I.; Megahed, T.F.; Nassef, M.G.A. Mitigation of Electrical Discharge Damage in Electric Vehicle Bearings: Comparative Study of Multi-Walled Carbon Nanotubes and Alumina Nanoparticles in Lubricating Grease. *Vehicles* **2025**, *7*, 19. [CrossRef]
47. United Nations. Sustainable Consumption and Production. United Nations Sustainable Development Goals Knowledge Platform. Available online: https://sdgs.un.org/topics/sustainable-consumption-and-production (accessed on 14 February 2025).
48. Sharma, N.; Bari, S.K.; Nagendramma, P.; Thakre, G.; Ray, A. Tribological Investigations of Sustainable Bio-Based Lubricants for Industrial Applications. In *Green Tribology*; Kasolang, S.B., Xie, G., Katiyar, J.K., Rani, A.M., Eds.; CRC Press: Boca Raton, FL, USA, 2021; pp. 45–72. Available online: https://www.taylorfrancis.com/chapters/edit/10.1201/9781003139386-3/tribological-investigations-sustainable-bio-based-lubricants-industrial-applications-neha-sharma-sayed-khadija-bari-ponnekanti-nagendramma-gananath-thakre-anjan-ray (accessed on 15 March 2025).
49. Nassef, B.G.; Pape, F.; Poll, G. Enhancing the Performance of Rapeseed Oil Lubricant for Machinery Component Applications through Hybrid Blends of Nanoadditives. *Lubricants* **2023**, *11*, 479. [CrossRef]
50. Syahrullaila, S.; Kamitani, S.; Shakirin, A. Tribological Evaluation of Mineral Oil and Vegetable Oil as a Lubricant. *Jurnal Teknologi* **2014**, *66*, 43–48. [CrossRef]

51. Habibullah, M.; Masjuki, H.H.; Kalam, M.A.; Ashraful, A.M.; Habib, M.A.; Mobarak, H.M. Effect of Bio-Lubricant on Tribological Characteristics of Steel. In Proceedings of the 10th International Conference on Mechanical Engineering (ICME 2013), Dhaka, Bangladesh, 20–22 December 2013; Procedia Engineering; Volume 90, pp. 740–745. [CrossRef]
52. Agrawal, S.M.; Lahane, S.; Patil, N.G.; Brahmankar, P.K. Experimental Investigations into Wear Characteristics of M2 Steel Using Cotton Seed Oil. *Procedia Eng.* **2014**, *97*, 4–14. [CrossRef]
53. Barbera, E.; Hirayama, K.; Maglinao, R.L.; Davis, R.W.; Kumar, S. Recent Developments in Synthesizing Biolubricants—A Review. *Biomass Conv. Biorefinery* **2024**, *14*, 2867–2887. [CrossRef]
54. Woma, T.Y. Vegetable Oil Based Lubricants: Challenges and Prospects. *Tribol. Online* **2019**, *14*, 60–69. [CrossRef]
55. Dube, N.N.; ElKady, M.; Noby, H.; Nassef, M.G.A. Developing a Sustainable Grease from Jojoba Oil with Plant-Waste-Based Nanoadditives for Enhancement of Rolling Bearing Performance. *Sci. Rep.* **2024**, *14*, 539. [CrossRef] [PubMed]
56. Masudi, A.; Muraza, O. Vegetable Oil to Biolubricants: Review on Advanced Porous Catalysts. *Energy Fuels* **2018**, *32*, 10295–10310. [CrossRef]
57. Masjuki, H.H. Palm Oil and Mineral Oil-Based Lubricants—Their Tribological and Emission Performance. *Tribol. Int.* **1999**, *32*, 209–215. [CrossRef]
58. Sukirno, R.F.; Bismo, S.; Nasikin, M. Biogrease Based on Palm Oil and Lithium Soap Thickener: Evaluation of Antiwear Property. *World Appl. Sci. J.* **2009**, *6*, 401–407.
59. Hassan, M.Y.; Ani, F.N.; Syahrullail, S. Tribological Performance of Refined, Bleached and Deodorised Palm Olein Blends as Bio-Lubricants. *J. Oil Palm Res.* **2016**, *28*, 510–519. [CrossRef]
60. Syahrullail, S.; Yahya Wira, J.; Wan Nik, W.B.; Fawwaz, W.N. Friction Characteristics of RBD Palm Olein Using Four-Ball Tribotester. *Appl. Mech. Mater.* **2013**, *315*, 936–940. [CrossRef]
61. Nassef, B.G.; Moradi, A.; Bayer, G.; Pape, F.; Abouelkasem, Z.A.; Rummel, F.; Schmölzer, S.; Poll, G.; Marian, M. Biogenic Palm Oil-Based Greases with Glycerol Monostearate and Soy Wax: A Rheological and Tribological Study. *Results Eng.* **2025**, *25*, 103728. [CrossRef]
62. Nassef, M.G.A.; Nassef, B.G.; Hassan, H.S.; Nassef, G.A.; Elkady, M.; Pape, F. Tribological and Chemical–Physical Behavior of a Novel Palm Grease Blended with Zinc Oxide and Reduced Graphene Oxide Nano-Additives. *Lubricants* **2024**, *12*, 191. [CrossRef]
63. Est-Aegis. How Do VFDs Cause Bearing Damage? Est-Aegis Blog. Available online: https://blog.est-aegis.com/how-do-vfds-cause-bearing-damage (accessed on 14 February 2025).
64. Pei, T.; Zhang, H.; Hua, W.; Zhang, F. Comprehensive Review of Bearing Currents in Electrical Machines: Mechanisms, Impacts, and Mitigation Techniques. *Energies* **2025**, *18*, 517. [CrossRef]
65. Tong, W. *Mechanical Design and Manufacturing of Electric Motors*, 2nd ed.; CRC Press: Boca Raton, FL, USA, 2022; Chapter 6. [CrossRef]
66. Farfan-Cabrera, L.I. Tribology of Electric Vehicles: A Review of Critical Components, Current State and Future Improvement Trends. *Tribol. Int.* **2019**, *138*, 473–486. [CrossRef]
67. Tawfiq, K.B.; Güleç, M.; Sergeant, P. Bearing Current and Shaft Voltage in Electrical Machines: A Comprehensive Research Review. *Machines* **2023**, *11*, 550. [CrossRef]
68. Romanenko, A. Study of Inverter-Induced Bearing Damage Monitoring in Variable-Speed-Driven Motor Systems. Doctoral Dissertation, Lappeenranta University of Technology, Lappeenranta, Finland, December 2017. Acta Universitatis Lappeenrantaensis, ISBN 978-952-335-185-1. Available online: https://lutpub.lut.fi/handle/10024/147584 (accessed on 20 January 2025).
69. Yousuf, A.; Spikes, H.; Guo, L.; Kadiric, A. Influence of Electric Potentials on Surface Damage in Rolling–Sliding Contacts Under Mixed Lubrication. *Tribol. Lett.* **2025**, *73*, 45. [CrossRef]
70. Êvo, M.T.A.; Alzamora, A.M.; Zaparoli, I.O.; de Paula, H. Inverter-Induced Bearing Currents: A Thorough Study of the Cause-and-Effect Chains. *IEEE Ind. Appl. Mag.* **2022**, *29*, 2–12. [CrossRef]
71. Zhu, X.; Shi, N.; Li, Y.; Yang, Y.; Wang, X. Bearing Capacitance Estimation of Electric Vehicle Driving Motor. *IOP Conf. Ser. Earth Environ. Sci.* **2018**, *199*, 032065. [CrossRef]
72. NSK Ltd. *Rolling Bearings Catalog*; (CAT No. E1102m); NSK Ltd.: Tokyo, Japan, 2013. Available online: https://www.nsk.com/content/dam/nsk/common/catalogs/ctrgPdf/e1102m.pdf (accessed on 29 March 2025).
73. Nik Roselina, N.R.; Mohamad, N.S.; Kasolang, S. Evaluation of TiO_2 Nanoparticles as Viscosity Modifier in Palm Oil Bio-lubricant. *IOP Conf. Ser. Mater. Sci. Eng.* **2020**, *834*, 012032. [CrossRef]
74. Nair, S.S.; Nair, K.P.; Rajendrakumar, P.K. Evaluation of Physicochemical, Thermal and Tribological Properties of Sesame Oil (*Sesamum indicum* L.): A Potential Agricultural Crop Base Stock for Eco-Friendly Industrial Lubricants. *Int. J. Agric. Resour. Governance Ecol.* **2017**, *13*, 77–90. [CrossRef]
75. Ngatirah, N.; Kusumastuti, K.; Tarigan, C.W.; Suparyanto, T.; Trinugroho, J.P.; Pardamean, B. Glycerolysis of Palm Kernel Oil Catalyzed by MgO on Mono and Diglyceride Composition and Their Antibacterial Activity. *BIO Web Conf.* **2024**, *94*, 02009. [CrossRef]

76. Faridah, D.N.; Sumitra, N.R.; Hariyadi, P.; Triana, R.N.; Laksana, A.J.; Andarwulan, N. Laboratory-Scale Synthesis of Mono-Diacylglycerol from Palm Oil Stearin Using Glycerolysis. In Proceedings of the 2nd SEAFAST International Seminar (2nd SIS 2019), Yogyakarta, Indonesia, 25–26 November 2019; SciTePress: Setúbal, Portugal, 2019. Available online: https://www.scitepress.org/Papers/2019/99777/99777.pdf (accessed on 21 May 2025).
77. *ASTM D445-24*; Standard Test Method for Kinematic Viscosity of Transparent and Opaque Liquids (and Calculation of Dynamic Viscosity). ASTM International: West Conshohocken, PA, USA, 2024.
78. *ASTM D217-21a*; Standard Test Methods for Cone Penetration of Lubricating Grease. ASTM International: West Conshohocken, PA, USA, 2021. Available online: https://store.astm.org/d0217-21a.html (accessed on 2 January 2025).
79. Lugt, P.M. *Grease Lubrication in Rolling Bearings*; Wiley: West Sussex, UK, 2013. [CrossRef]
80. *ASTM D2265-22*; Standard Test Method for Dropping Point of Lubricating Grease Over Wide Temperature Range. ASTM International: West Conshohocken, PA, USA, 2022. Available online: https://store.astm.org/d2265-22.html (accessed on 20 April 2025).
81. *ASTM D1742-20*; Standard Test Method for Oil Separation from Lubricating Grease During Storage. ASTM International: West Conshohocken, PA, USA, 2020. [CrossRef]
82. *ASTM G99-17*; Standard Test Method for Wear Testing with a Pin-on-Disk Apparatus. ASTM International: West Conshohocken, PA, USA, 2017. Available online: https://store.astm.org/g0099-17.html (accessed on 7 February 2025).
83. Cañellas, G.; Emeric, A.; Combarros, M.; Navarro, A.; Beltran, L.; Vilaseca, M.; Vives, J. Tribological Performance of Esters, Friction Modifier and Antiwear Additives for Electric Vehicle Applications. *Lubricants* **2023**, *11*, 109. [CrossRef]
84. Bechev, D.; Capan, R.; Gonda, A.; Sauer, B. Method for the Investigation of the EDM Breakdown Voltage of Grease and Oil on Rolling Bearings. *Bear. World J.* **2019**, *4*, 83–91.
85. Joshi, A.; Blennow, J. Investigation of the Static Breakdown Voltage of the Lubricating Film in a Mechanical Ball Bearing. In Proceedings of the Nordic Insulation Symposium (NORDIS), Trondheim, Norway, 9–12 June 2013; pp. 1–8. [CrossRef]
86. Beroual, A.; Khaled, U.; Mbolo Noah, P.S.; Sitorus, H. Comparative Study of Breakdown Voltage of Mineral, Synthetic and Natural Oils and Based Mineral Oil Mixtures under AC and DC Voltages. *Energies* **2017**, *10*, 511. [CrossRef]
87. Gonda, A.; Capan, R.; Bechev, D.; Sauer, B. The Influence of Lubricant Conductivity on Bearing Currents in the Case of Rolling Bearing Greases. *Lubricants* **2019**, *7*, 108. [CrossRef]
88. Graf, S.; Koch, O.; Sauer, B. Influence of Parasitic Electric Currents on an Exemplary Mineral-Oil-Based Lubricant and the Raceway Surfaces of Thrust Bearings. *Lubricants* **2023**, *11*, 313. [CrossRef]
89. Shen, T.; Wang, D.; Yun, J.; Liu, Q.; Liu, X.; Peng, Z. Mechanical Stability and Rheology of Lithium–Calcium-Based Grease Containing ZDDP. *RSC Adv.* **2016**, *6*, 11637–11647. [CrossRef]
90. *ASTM D1831-21*; Standard Test Method for Measuring Oxidation Stability of Automatic Transmission Fluids (Rotating Pressure Vessel Oxidation Test—RPVOT). ASTM International: West Conshohocken, PA, USA, 2021. Available online: https://store.astm.org/d1831-21.html (accessed on 20 March 2025).
91. Tan, K.-H.; Awala, H.; Mukti, R.R.; Wong, K.-L.; Ling, T.C.; Mintova, S.; Ng, E.-P. Zeolite Nanoparticles as Effective Antioxidant Additive for the Preservation of Palm Oil-Based Lubricant. *J. Taiwan Inst. Chem. Eng.* **2016**, *58*, 565–571. [CrossRef]
92. *ASTM D7346-15R21*; Standard Test Method for No Flow Point and Pour Point of Petroleum Products and Liquid Fuels. ASTM International: West Conshohocken, PA, USA, 2021. [CrossRef]
93. Hamrock, B.J.; Dowson, D. Isothermal Elastohydrodynamic Lubrication of Point Contacts: Part 1—Theoretical Formulation. *ASME J. Lubr. Technol.* **1976**, *98*, 223–228. [CrossRef]
94. Hamrock, B.J.; Dowson, D.; Tallian, T.E. Ball Bearing Lubrication (The Elastohydrodynamics of Elliptical Contacts). *ASME J. Lubr. Technol.* **1982**, *104*, 279–281. [CrossRef]
95. Biresaw, G.; Bantchev, G.B. Pressure Viscosity Coefficient of Vegetable Oils. *Tribol. Lett.* **2013**, *49*, 501–512. [CrossRef]
96. Maruyama, T.; Radzi, F.; Sato, T.; Iwase, S.; Maeda, M.; Nakano, K. Lubrication Condition Monitoring in EHD Line Contacts of Thrust Needle Roller Bearing Using the Electrical Impedance Method. *Lubricants* **2023**, *11*, 223. [CrossRef]
97. Paredes, X.; Comuñas, M.J.P.; Pensado, A.S.; Bazile, J.-P.; Boned, C.; Fernández, J. High Pressure Viscosity Characterization of Four Vegetable and Mineral Hydraulic Oils. *Ind. Crops Prod.* **2014**, *54*, 281–290. [CrossRef]
98. Lotfizadeh Dehkordi, B.; Shiller, P.J.; Doll, G.L. Pressure- and Temperature-Dependent Viscosity Measurements of Lubricants with Polymeric Viscosity Modifiers. *Front. Mech. Eng.* **2019**, *5*, 18. [CrossRef]
99. Cen, H.; Lugt, P.M.; Morales-Espejel, G.E. On the Film Thickness of Grease-Lubricated Contacts at Low Speed. *Tribol. Trans.* **2014**, *57*, 668–678. [CrossRef]
100. Cen, H.; Lugt, P.M.; Morales-Espejel, G.E. Film Thickness of Mechanically Worked Lubricating Grease at Very Low Speeds. *Tribol. Trans.* **2014**, *57*, 1066–1071. [CrossRef]
101. Cen, H.; Lugt, P.M. Film Thickness in a Grease Lubricated Ball Bearing. *Tribol. Int.* **2019**, *134*, 26–35. [CrossRef]
102. Hamrock, B.J.; Schmid, S.R.; Jacobson, B.O. *Fundamentals of Fluid Film Lubrication*, 2nd ed.; CRC Press: Boca Raton, FL, USA, 2004. [CrossRef]

103. Gropper, D.; Wang, L.; Harvey, T.J. Hydrodynamic Lubrication of Textured Surfaces: A Review of Modeling Techniques and Key Findings. *Tribol. Int.* **2016**, *94*, 509–529. [CrossRef]
104. Garg, P.; Kumar, A.; Thakre, G.; Jain, A.K. Influence of Temperature on the Tribological Behavior of Bio-Ester Based Lubricant. *J. Comput. Theor. Nanosci.* **2016**, *22*, 3805–3809. [CrossRef]
105. Schneider, V.; Bader, N.; Liu, H.; Poll, G. Method for *In Situ* Film Thickness Measurement of Ball Bearings under Combined Loading Using Capacitance Measurements. *Tribol. Int.* **2022**, *171*, 107524. [CrossRef]
106. Bader, N.; Furtmann, A.; Tischmacher, H.; Poll, G. Capacitances and Lubricant Film Thicknesses of Grease and Oil Lubricated Bearings. In Proceedings of the 72nd STLE Annual Meeting & Exhibition, Rolling Element Bearings II, Atlanta, GA, USA, 21–25 May 2017. Available online: https://www.stle.org/images/pdf/STLE_ORG/AM2017%20Presentations/Rolling%20Element%20Bearings/STLE2017_Rolling%20Element%20Bearings%20II_Session%206C_N.%20Bader_Capacitances%20and%20Lubricant%20Film%20Thicknesses%20of%20Grease%20and%20Oil%20Lubricated%20Bearings.pdf (accessed on 21 February 2025).
107. Maruyama, T.; Maeda, M.; Nakano, K. Lubrication Condition Monitoring of Practical Ball Bearings by Electrical Impedance Method. *Tribol. Online* **2019**, *14*, 327–338. [CrossRef]
108. Joly-Pottuz, L.; Martin, J.M.; De Barros Bouchet, M.I.; Mogne, T.L.; Belin, M.; Montagnac, G.; Reynard, B.; Epicier, T.; Kano, M.; Enomoto, Y.; et al. Anomalous Low Friction Under Boundary Lubrication of Steel Surfaces by Polyols. *Tribol. Lett.* **2009**, *34*, 21–29. [CrossRef]
109. Ng, W.S.; Lee, C.S.; Cheng, S.-F.; Chuah, C.H.; Wong, S.F. Biocompatible Polyurethane Scaffolds Prepared from Glycerol Monostearate-Derived Polyester Polyol. *J. Polym. Environ.* **2018**, *26*, 2881–2900. [CrossRef]
110. Gulzar, M.; Masjuki, H.H.; Kalam, M.A.; Varman, M.; Zulkifli, N.W.M.; Mufti, R.A.; Zahid, R. Tribological Performance of Nanoparticles as Lubricating Oil Additives. *J. Nanopart. Res.* **2016**, *18*, 223. [CrossRef]
111. Mang, T.; Dresel, W. (Eds.) *Lubricants and Lubrication*; Wiley-VCH: Weinheim, Germany, 2007. [CrossRef]
112. Alves, S.M.; Barros, B.S.; Trajano, M.F.; Ribeiro, K.S.B.; Moura, E. Tribological behavior of vegetable oil-based lubricants with nanoparticles of oxides in boundary lubrication conditions. *Tribol. Int.* **2013**, *65*, 28–36. [CrossRef]
113. Rojacz, H.; Maierhofer, D.; Piringer, G. Environmental impact evaluation of wear protection materials. *Wear* **2025**, *560–561*, 205612. [CrossRef]
114. Nugroho, A.; Kozin, M.; Mamat, R.; Zhang, B.; Ghazali, M.F.; Kamil, M.P.; Puranto, P.; Fitriani, D.A.; Azahra, S.A.; Suwondo, K.P.; et al. Enhancing Tribological Performance of Electric Vehicle Lubricants: Nanoparticle-Enriched Palm Oil Biolubricants for Wear Resistance. *Heliyon* **2024**, *10*, e39742. [CrossRef]
115. Vyavhare, K.; Aswath, P.B. Tribological Properties of Novel Multi-Walled Carbon Nanotubes and Phosphorus-Containing Ionic Liquid Hybrids in Grease. *Front. Mech. Eng.* **2019**, *5*, 15. [CrossRef]
116. Guo, L.; Mol, H.; Nijdam, T.; de Vries, L.; Bongaerts, J. Study on the Electric Discharge Behaviour of a Single Contact in EV Motor Bearings. *Tribol. Int.* **2023**, *187*, 108743. [CrossRef]
117. Schneider, V.; Krewer, M.; Poll, G.; Marian, M. Effect of Harmful Bearing Currents on the Service Life of Rolling Bearings: From Experimental Investigations to a Predictive Model. *Lubricants* **2024**, *12*, 230. [CrossRef]
118. Muangpratoom, P.; Suriyasakulpong, C.; Maneerot, S.; Vittayakorn, W.; Pattanadech, N. Experimental Study of the Electrical and Physiochemical Properties of Different Types of Crude Palm Oils as Dielectric Insulating Fluids in Transformers. *Sustainability* **2023**, *15*, 14269. [CrossRef]
119. Slita, A.; Filali, N.; El Hammari, L.; Hassanain, I.; Sifou, A.; Saoiabi, S.; El Belghiti, M.A. Study of the Electrical Resistivity of Several Vegetable Oils: Nigella, Olive, Argan, Prickly Pear and Palm. *Pharm. Lett.* **2016**, *8*, 62–64.
120. Emebu, S.; Osaikhuiwuomwan, O.; Mankonen, A.; Udoye, C.; Okieimen, C.; Janáčová, D. Influence of Moisture Content, Temperature, and Time on Free Fatty Acid in Stored Crude Palm Oil. *Sci. Rep.* **2022**, *12*, 9846. [CrossRef] [PubMed]
121. Hossain, M.K.; Khan, M.A.G. An Auspicious Dielectric in High Voltage Engineering: Vegetable Oil. In Proceedings of the 2015 International Conference on Electrical & Electronic Engineering (ICEEE), Rajshahi, Bangladesh, 4–6 November 2015; IEEE: Piscataway, NJ, USA, 2016; pp. 153–156. [CrossRef]
122. Rajab, A.; Sulaeman, A.; Sudirham, S.; Suwarno, S. A Comparison of Dielectric Properties of Palm Oil with Mineral and Synthetic Types Insulating Liquid under Temperature Variation. *J. Eng. Technol. Sci.* **2011**, *43*, 191–208. [CrossRef]
123. Singh, S.P.; Kumar, P.; Manohar, R.; Shukla, J.P. Dielectric Properties of Some Oil Seeds at Different Concentration of Moisture Contents and Micro-fertilizer. *Int. J. Agric. Res.* **2006**, *1*, 293–304. [CrossRef]
124. *ISO 13373-2:2016*; Condition Monitoring and Diagnostics of Machines—Vibration Condition Monitoring—Part 2: Processing, Analysis and Presentation of Vibration Data, 2nd ed. ISO: Geneva, Switzerland, 2016. Available online: https://www.iso.org/standard/68128.html (accessed on 10 January 2025).
125. *ISO 10816-7:2009*; Mechanical Vibration—Evaluation of Machine Vibration by Measurements on Non-Rotating Parts—Part 7: Rotodynamic Pumps for Industrial Applications, Including Measurements on Rotating Shafts, 1st ed. ISO: Geneva, Switzerland, 2009. Available online: https://www.iso.org/standard/41726.html (accessed on 15 February 2025).

126. Feng, Z.; Ma, H.; Zuo, M.J. Spectral Negentropy Based Sidebands and Demodulation Analysis for Planet Bearing Fault Diagnosis. *J. Sound Vib.* **2017**, *410*, 124–150. [CrossRef]
127. Albezzawy, M.N.; Nassef, M.G.A.; Elsayed, E.S.; Elkhatib, A. Early Rolling Bearing Fault Detection Using A Gini Index Guided Adaptive Morlet Wavelet Filter. In Proceedings of the 2019 IEEE 10th International Conference on Mechanical and Aerospace Engineering (ICMAE), Brussels, Belgium, 22–25 July 2019; pp. 314–322. [CrossRef]
128. Reddy, A.B.; Shah, F.U.; Leckner, J.; Rutland, M.W.; Glavatskih, S. Ionic Liquids Enhance Electrical Conductivity of Greases: An Impedance Spectroscopy Study. *Colloids Surf. A Physicochem. Eng. Asp.* **2024**, *683*, 132875. [CrossRef]
129. Prashad, H. Appearance of Craters on Track Surface of Rolling Element Bearings by Spark Erosion. *Tribol. Int.* **2001**, *34*, 39–47. [CrossRef]
130. Kostal, D.; Sperka, P.; Chmelar, J.; Vitek, P.; Polak, M.; Krupka, I. Comparison of Grease and Solid Lubrication of Rolling Bearings under Small-Stroke Reciprocation for Space Applications. *Tribol. Ind.* **2020**, *42*, 146–158. [CrossRef]
131. Chen, Y.; Peng, K.; Li, X.; Sun, L.; Su, T.; Zhang, W.; Zhang, D.; Fan, S.; Yin, H.; Zhou, M. Lubricating and Conductive Properties of Modified Graphene/Silver Nanoparticles Under Current-Carrying Friction Conditions. *Lubricants* **2025**, *13*, 38. [CrossRef]
132. Jason, Y.J.J.; How, H.G.; Teoh, Y.H.; Chuah, H.G. A Study on the Tribological Performance of Nanolubricants. *Processes* **2020**, *8*, 1372. [CrossRef]
133. Bhaumik, S.; Prabhu, S.; Singh, K.J. Analysis of Tribological Behavior of Carbon Nanotube-Based Industrial Mineral Gear Oil (250 cSt Viscosity). *Adv. Tribol.* **2014**, *2014*, 341365. [CrossRef]
134. *ASTM D3336-20*; Standard Test Method for Life of Lubricating Greases in Ball Bearings at Elevated Temperatures. ASTM International: West Conshohocken, PA, USA, 2020. Available online: https://store.astm.org/d3336-20.html (accessed on 20 June 2025).

MDPI AG
Grosspeteranlage 5
4052 Basel
Switzerland
Tel.: +41 61 683 77 34

Lubricants Editorial Office
E-mail: lubricants@mdpi.com
www.mdpi.com/journal/lubricants

www.ingramcontent.com/pod-product-compliance
Lightning Source LLC
Chambersburg PA
CBHW062032220926
51850CB00010B/2520